31 Advances in Polymer Science

Fortschritte der Hochpolymeren-Forschung

Edited by H.-J. Cantow, Freiburg i. Br. · G. Dall'Asta, Colleferro
K. Dušek, Prague · J. D. Ferry, Madison · H. Fujita, Osaka
M. Gordon, Colchester · W. Kern, Mainz · G. Natta, Milano
S. Okamura, Kyoto · C. G. Overberger, Ann Arbor · T. Saegusa, Kyoto
G. V. Schulz, Mainz · W. P. Slichter, Murray Hill · J. K. Stille, Fort Collins

With 45 Figures

Springer-Verlag
Berlin Heidelberg GmbH **1979**

Editors

ISBN 978-3-662-15428-1   ISBN 978-3-540-35391-1 (eBook)
DOI 10.1007/978-3-540-35391-1

Library of Congress Catalog Card Number 61-642

Contents

Stereospecific Polymerization of
Alpha-Substituted Acrylic Acid Esters

Heimei Yuki and Koichi Hatada

Department of Chemistry, Faculty of Engineering Science, Osaka University, Toyonaka, Osaka, Japan

A large number of methacrylates containing various ester groups have been polymerized and copolymerized with radical and anionic initiators, such as organoalkali metal, magnesium and aluminum compounds. These polymerization and copolymerization are reviewed with special reference on their stereospecificity. The polymerizations of α-phenyl, α-alkyl and α-chloroacrylates are also included.

Table of Contents

1 Introduction

Since the first preparation of stereoregular poly(methyl methacrylate) by Fox et al.[1] and Miller et al.[2] in 1958, a large number of papers have been published on the stereospecific polymerization of methyl methacrylate, while the NMR technique for the determination of microstructure developed by Bovey and Tiers[3] and Nishioka et al.[4] enabled us to accumulate the extensive information on this polymerization. Mostly anionic initiators have been used for the polymerization. A review on the polymerization by lithium compounds was presented by Bywater[5]. In a recent review by Pino and Suter[6] were discussed some of the factors which can influence the stereoregulation in the polymerization of vinyl monomers including α-substituted acrylate. A variety of magnesium and aluminum compounds can be utilized as stereospecific initiators. Besides methyl methacrylate, not only methacrylates with various ester groups, but also α-substituted acrylates, such as α-ethyl- or α-phenyl-acrylate, were also subjected to the stereospecific polymerization by anionic initiator. The stereospecificity in the copolymerization between the monomers described above is also a matter of interest.

This article will mainly deal with the stereospecific polymerization and copolymerization of these monomers rather than the polymerization of methyl methacrylate itself. The association of polymer is one of the characteristic features of poly-(methyl methacrylate) in solution. The influence of this "stereo-complex" formation on the stereospecificity in propagation step should not be ignored.

2 Reproducibility in Polymerization of Methyl Methacrylate

The data on the anionic polymerization of methyl methacrylate are widely varied among the literatures reported. Not only the microstructure but also the molecular weight and even the yield of polymer vary with investigator, even if the same reaction conditions have been employed.[7, 8] When the polymerization was done by five graduate students under the same conditions using free radical initiator, the microstructure of the produced polymers was close to each other and the differences among them were nearly the same level of the fluctuation in NMR measurement[8] (Table 1). The error of the NMR measurement is within 5 % and the personal error in the measurement can be neglected, if the instrument is properly operated.[9, 10] On the other hand, the data obtained in the polymerization with BuLi by the students were much more scattered. However, if the same student performed five runs of the experiments carefully, the results coincided with each other within the error of the NMR measurement as shown in Table 1. One of the major causes of the variation in the anionic polymerization seems to be the contamination by impurity such as a trace of water.[11, 12] However, the rate of stirring at the mixing of an initiator and a monomer solution cannot be neglected, even if the amount of impurity can be controlled at a constant level. This is a matter of course, since the concentration of effective active species may be dominated by extremely fast initiation accompanied with competitive side reactions.[13-17] The efficiency of initiator is strongly affect-

Table 1. Polymerization of methyl methacrylate by five students[a], Ref. [8]

No.	Polymerization	Yield (%)	$[\eta]^b$ (dl/g)	Tacticity (%)		
				I	H	S
1	AIBN, Toluene, 60 °C, 48 hr	91.5 (3.1)	0.26 (7.6)	5.9 (6.0)	35.7 (1.8)	58.4 (1.4)
2A	BuLi, Toluene, −78 °C, 24 hr	63.6 (56.4)	0.62 (31.5)	71.6 (8.4)	17.5 (25.0)	10.9 (18.0)
2B[c]	BuLi, Toluene, −78 °C, 24 hr	77.4 (2.5)	0.44 (5.8)	72.3 (1.9)	16.5 (5.7)	11.2 (9.0)
3A	BuLi, THF, −78 °C, 24 hr	66.4 (60.1)	0.59 (38.7)	6.5 (14.2)	37.0 (4.9)	56.6 (3.8)
3B[c]	BuLi, THF, −78 °C, 24 hr	87.2 (3.8)	0.64 (7.5)	5.7 (5.1)	37.7 (2.5)	56.7 (1.9)

[a] The figure in parenthesis represents the precision (%).
[b] Determined in toluene at 30.0 ± 0.03 °C.
[c] The results from five runs done by one student.

ed by the absolute concentrations of monomer and initiator as well as the relative ones in the anionic polymerization of methyl methacrylate. The side reactions proceed much more with the higher concentration of BuLi and with the smaller ratio of monomer to initiator concentration.[18, 19] So the differences in the mixing rate of initiator and monomer should cause the fluctuation of the results of the polymerizations. When the polymerization by BuLi in toluene was performed without agitation, it gave a polymer with markedly high molecular weight and isotacticity in a high yield, compared with the conventional procedure. This is an extreme example and called "slow growth polymerization", which will be described later.[20, 21] Amerik et al.[22] obtained poly(methyl methacrylate) with high isotacticity by adding the monomer in vapor phase on to the surface of initiator solution in a vacuum system. They explained that the high isotacticity was resulted by the low polarity of the polymerization system. Vapor phase addition of monomer has been also used to get a rapid and completely uniform way of mixing, which results in the formation of monodispersed poly(methyl methacrylate).[23, 24]

3 Determination of Stereoregularity in Poly(alkyl methacrylate) other than Poly(methyl methacrylate)

In the PMR spectrum of poly(alkyl methacrylate) other than poly(methyl methacrylate) the resonance of an ester group often overlaps with the α-methyl signal and obscures its splitting due to the tacticity of the polymer. Therefore, the measurement of the tacticity was usually done after the polymer was converted to poly-(methyl methacrylate) through hydrolysis by sulfuric acid and then methylation with diazomethane.[25] Recently it was found that the magnitude of spin-lattice relaxation time T_1 of protons in poly(alkyl methacrylate) was in the order: backbone methylene $<$ α-methyl $\ll$ ester group.[26–29] This large difference in T_1 of α-methyl and ester protons was utilized to eliminate the ester-alkyl resonance overlapping with α-methyl signal, and it enabled us to determine the triad tacticity directly from the resonance of α-methyl protons in the original polymer.[30] The ester group of benzyl derivative is quantitatively eliminated by hydrogen bromide in an organic solvent such as toluene. Furthermore, tertiary ester such as t-butyl, α,α-dimethylbenzyl and trityl esters can be easily hydrolyzed by reflux in methanol containing a few drops of hydrochloric acid.[31–33] By these reactions we can not only hydrolyze the ester groups under milder conditions but also selectively eliminate only a tertiary or a benzylic ester group, if it is contained in one of the component monomers in a copolymer.[33–35] Poly(phenyl methacrylate) and poly(naphthyl methacrylate) are easily converted to poly(methyl methacrylate) by treatment with sodium methoxide in dimethyl sulfoxide.[36]

4 Stereoregularity of Polymethacrylates Prepared with Radical Initiators

Table 2 shows the microstructure of poly(methacrylic ester)s which were prepared with radical initiators. Most methacrylates gave rather syndiotactic polymers. How-

Table 2. Stereoregularity of polymethacrylates obtained by radical initiators

Methacrylate	Initiator	Temp. (°C)	Tacticity (%)			Ref.
			I	H	S	
Methyl	BPO	60	4	34	62	37)
Ethyl	BPO	70	8	23	69	38)
Isopropyl	BPO	70	7	31	62	38)
Butyl	AIBN	70	8	27	65	38)
Sec-butyl	AIBN	60	6	36	58	37)
Tert-butyl	BPO	70	8	40	52	38)
3-Pentyl	BPO	60	11	46	43	37)
1,1-Diethylpropyl	AIBN	50	14	53	33	37)
4-(2,6-Dimethyl)-heptyl	BPO	60	15	50	35	37)
Cyclopentyl	BPO	60	5	35	60	39)
Cyclohexyl	BPO	70	7	37	56	37)
4-Methylcyclohexyl	BPO	60	7	39	54	37)
3-Methylcyclohexyl	BPO	60	9	40	51	37)
2-Methylcyclohexyl	BPO	60	10	43	47	37)
4-Tert-butylcyclohexyl	AIBN	50	11	35	54	40)
D-Bornyl	AIBN	70	9	36	55	38)
DL-Isobornyl	AIBN	70	13	32	55	38)
l-Menthyl	BPO	60	13	47	40	37)
dl-Menthyl	BPO	60	13	47	40	41)
Benzyl	BPO	60	7	37	56	37)
DL-α-Methylbenzyl	AIBN	60	6	36	58	37)
Diphenylmethyl	AIBN	60	2	41	57	32)
α,α-Dimethylbenzyl	AIBN	60	11	47	42	34)
1,1-Diphenylethyl	AIBN	60	19	49	32	42)
Trityl	AIBN	60	64	22	14	32)
p-Phenylbenzyl	AIBN	80	5	38	58	43)
2,4,6-Triphenylbenzyl	AIBN	80	16	33	51	38)

2-Phenylethyl	BPO	60	7	37	56	37)
Phenyl	AIBN	80	3	36	61	44)
2-Methoxyphenyl	AIBN	80	6	41	53	44)
4-Methoxyphenyl	AIBN	80	10	32	58	44)
4-Tert-butylphenyl	AIBN	50	14	40	46	40)
1-Naphthyl[a]	AIBN	60	34	40	26	39)
2-Naphthyl[a]	AIBN	60	12	38	50	39)
2-Naphthyl[b]	AIBN	60	58	24	18	39)
5,6,7,8-Tetrahydronaphthyl	AIBN	80	12	46	42	39)
Decahydronaphthyl	AIBN	50	14	37	49	40)
Decahydronaphthyl	AIBN	60	6	33	61	39)
9-Fluorenyl	BPO	80	10	41	49	39)
Dimethyl-6,6-tricyclo[6,4,0,0^{2,7}]-dodecyl-3	AIBN	80	12	38	50	45)
2-Epoxypropyl	Diisopropylperoxydicarbonate-dimethylaniline	60	3	37	60	46)
2-Epoxypropyl	Diisopropylperoxydicarbonate-dimethylaniline	−78	0	5	95	46)

a Polymerized in benzene.
b Polymerized in hexane.

ever, the syndiotacticity decreased with increasing bulkiness of monomers having secondary and tertiary ester groups.[38, 40] It is accompanied with the increase of isotacticity, and trityl methacrylate gave an isotactic polymer even by radical initiator.[31, 32] Matsuzaki et al. found a correlation between the stereoregularity of some polymethacrylates and Taft's steric factor (Es) of the ester groups.[37] It was reported that the aromatic polymethacrylates have higher isotacticity and lower syndiotacticity than the corresponding saturated polymers (Table 2). This was explained by specific interaction in the polymerization between the aromatic rings.[39, 40] It should be also noted that 2-naphthyl methacrylate gave an isotactic polymer in hexane while it yielded a syndiotactic polymer in benzene. 1-Naphthyl methacrylate formed an atactic polymer independently of the solvent used.[39] Akashi et al.[47] observed that the microstructure was not unusual in radically prepared polymethacrylates which contained purine bases in the ester groups.

5 Stereoregularity of Polymethacrylates Prepared with Alkyllithium in Toluene and in Tetrahydrofuran

The triad tacticity of polymers produced by BuLi in toluene is summarized in Table 3. Isotactic polymers were formed from most monomers having hydrocarbon ester groups as known in the polymerization of methyl methacrylate. The highest isotacticity was observed in the polymers of cyclopentylmethyl, diphenylmethyl and trityl methacrylates. However, surprisingly 1,1-diphenylethyl methacrylate gave a rather syndiotactic polymer in this nonpolar solvent at −78 °C, in spite of its bulkiness which is similar to those of diphenylmethyl and trityl methacrylates, although it formed an isotactic polymer at 0 °C.[42] It is noticeable that the isotacticity in the polymer of optically active α-methylbenzyl methacrylate was higher than that of the polymer from the racemic monomer.[49] The same phenomenon was observed in the polymerization of menthyl methacrylate by PhMgBr and by LiAlH₄.[41]

Fowells et al.[48] polymerized stereospecifically deuterated acrylic and methacrylic esters to highly isotactic polymers under several conditions with a number of anionic initiators. They analyzed the polymers by ^{1}H NMR spectroscopy to get the information on the mode of monomer approach to the growing anion. The results of the polymerization initiated with fluorenyllithium suggest that there are two transition states leading to an isotactic placement which differ in the degree of solvation. One is derived from a naked contact ion pair and forms the polymer of threo-meso configuration. The other is apparently formed by the specifically solvated contact ion pair and forms erythro-meso. A very similar study was reported on the anionic polymerization of deuterated methyl acrylate.[54−56]

Various methacrylates containing heteroatoms in the ester groups were polymerized by Iwakura et al.[50, 53] A wide variety of microstructures were observed in the resulted polymers from 96 % isotacticity to 66 % syndiotacticity. In general the syndiotacticity was found to depend upon the basicity of the ester group.

The stereoregularity of polymers prepared by BuLi in THF is shown in Table 4. At −78 °C, syndiotactic polymers were mainly obtained. Diphenylmethyl meth-

Table 3. Stereoregularity of polymethacrylates obtained by alkyllithium in toluene

Methacrylate	Initiator	Temp. (°C)	Tacticity (%)			Ref.
			I	H	S	
Methyl	BuLi	0	81	14	5	43)
Methyl	BuLi	−78	72	17	11	8)
Ethyl	BuLi	−70	82	10	8	38)
Ethyl	BuLi	−78	91	8	1	48)
Ethyl	DPHL[a]	−78	89	10	1	48)
Isopropyl	BuLi	−78	88	6	6	30)
Isopropyl	BuLi	−70	70	16	14	38)
Butyl	BuLi	−78	85	8	7	30)
Tert-butyl	BuLi	−70	90	5	5	38)
2-Methylbutyl	BuLi	−78	92	6	2	30)
Benzyl	BuLi	0	73	22	5	32)
Benzyl	BuLi	−78	81	15	4	32)
p-Phenylbenzyl	BuLi	0	76	20	4	43)
(R)-α-Methylbenzyl	BuLi	−78	78	17	5	49)
(RS)-α-Methylbenzyl	BuLi	0	70	24	6	49)
(RS)-α-Methylbenzyl	BuLi	−78	56	35	9	49)
Diphenylmethyl	BuLi	0	93	4	3	32)
Diphenylmethyl	BuLi	−78	99	1	0	32)
α,α-Dimethylbenzyl	BuLi	0	60	27	13	34)
α,α-Dimethylbenzyl	BuLi	−78	68	19	13	34)
1,1-Diphenylethyl	BuLi	0	52	37	11	42)
1,1-Diphenylethyl	BuLi	−78	23	28	49	42)
Trityl	BuLi	0	93	4	3	32)
Trityl	BuLi	−78	96	2	2	32)
2-Phenylethyl	BuLi	0	76	19	5	37)
Phenyl	BuLi	0	92	5	3	37)
Phenyl	FlLi[b]	0	82	9	9	44)

Table 3 (continued)

Methacrylate	Initiator	Temp. (°C)	Tacticity (%) I	Tacticity (%) H	Tacticity (%) S	Ref.
Phenyl	BuLi	−78	71	25	4	37)
4-Methoxyphenyl	FILi[b]	0	86	9	5	44)
Cyclopropanemethyl	BuLi	−78	92	4	4	50)
Cyclopentylmethyl	BuLi	−78	100	0	0	50)
Dimethyl-6,6-tricyclo[6,4,0,0^{2,7}]dodecyl-3	BuLi	0	74	22	4	45)
Trimethylsilyl	BuLi	−70	89	8	3	51)
Allyl	DPHL[a]	−80	90	10	0	52)
2-Methoxyethyl	BuLi	−78		34	66	50)
2-Methylthioethyl	BuLi	−78	96	4		50)
3-Methoxypropyl	BuLi	−78	27	38	35	50)
4-Methoxybutyl	BuLi	−78	59	22	19	50)
2,3-Epoxypropyl	BuLi	−78	33	38	29	53)
2,3-Epithiopropyl	BuLi	−78	95	2	3	53)
Tetrahydrofurfuryl	BuLi	−78	4	46	50	50)
Furfuryl	BuLi	−78	86	11	3	50)
N,N-Dimethylaminoethyl	BuLi	−78	73	16	11	50)
2-Pyridylmethyl	BuLi	−78	10	51	39	50)

a 1,1-Diphenylhexyllithium.
b Fluorenyllithium.

Table 4. Stereoregularity of polymethacrylates obtained by alkyllithium in THF

Methacrylate	Initiator	Temp. (°C)	Tacticity (%)			Ref.
			I	H	S	
Methyl	BuLi	0	31	32	37	43)
Methyl	BuLi	−78	6	38	56	43)
Ethyl	FlLi[a]	−78	5	21	74	48)
Ethyl	DPHL[b]	−78	7	40	53	48)
Isopropyl	BuLi	−78	13	32	55	30)
Butyl	BuLi	−78	4	29	67	30)
Allyl	DPHL[b]	−80	0	15	85	52)
Cyclopropanemethyl	BuLi	−78	7	46	47	53)
Benzyl	BuLi	0	18	33	49	32)
Benzyl	BuLi	−78	6	31	63	32)
p-Phenylbenzyl	BuLi	−78	9	30	61	43)
(R)-α-Methylbenzyl	BuLi	−78	12	28	60	49)
(RS)-α-Methylbenzyl	BuLi	0	16	39	45	49)
(RS)-α-Methylbenzyl	BuLi	−78	8	31	60	49)
Diphenylmethyl	BuLi	0	2	31	67	32)
Diphenylmethyl	BuLi	−78	2	11	87	32)
α,α-Dimethylbenzyl	BuLi	0	11	37	52	34)
α,α-Dimethylbenzyl	BuLi	−78	10	29	61	34)
1,1-Diphenylethyl	BuLi	0	17	48	35	42)
1,1-Diphenylethyl	BuLi	−78	21	46	33	42)
Trityl	BuLi	0	81	13	6	32)
Trityl	BuLi	−78	94	4	2	32)
Phenyl	FlLi[a]	−78	4	26	70	44)
2-Methoxyphenyl	FlLi[a]	−78	4	30	66	44)
4-Methoxyphenyl	FlLi[a]	−78	4	44	52	44)

[a] Fluorenyllithium.
[b] 1,1-Diphenylhexyllithium.

acrylate gave a polymer of especially high syndiotacticity. On the other hand trityl methacrylate formed a highly isotactic polymer not only in toluene as described already but also in this polar solvent. The large trityl ester group may prevent the syndiotactic placement of the incoming monomer to the growing chain end. From the observation of molecular model the polymer may take a helical structure which forces the isotactic addition at the chain end due to the rigid conformation allowed[31, 32)

    In the polymerization of methyl methacrylate in toluene the isotacticity of the polymer is usually decreased by the addition of a polar compound such as ether or amine. Only one third moles of N,N,N',N'-tetramethylethylenediamine against alkyllithium gave a rather syndiotactic polymer, while its methylene and trimethylene homologues merely reduced the isotacticity to a slight extent.[57] On the other hand, the increase of the isotacticity was observed in the polymerization with the addition of a small amount of aminoalcohol.[58]

6 Stereospecific Polymerization of Methacrylates with Alkali Metal Compounds other than Alkyllithium

Lithium tertiary alkoxide such as lithium tert-butoxide or 1,1-dimethylbutoxide can give high molecular poly(methyl methacrylate) even in toluene. The polymer obtained was highly isotactic.[59, 60] Addition of piperidine increased the initial rate of polymerization and decreased the isotacticity of the polymer with the concentration of piperidine but in small extent. This is contrary to the pronounced decrease in the isotacticity which has been observed in the polymerization by alkyllithium with the addition of a very small amount of similar electron-donating compound. This may suggest that the active growing center in tert-butoxide polymerization has more complex character and is much more stable than that in the alkyllithium polymerization.[61] In the polymerization of 2-methoxyethyl methacrylate by lithium tert-butoxide the marked change in the microstructure of the formed polymer was observed during the polymerization. It turned from predominantly isotactic first growth

Table 5. Stereoregularity of polymethacrylate prepared by alkali-metal compounds other than alkyllithium

Monomer	Initiator	Solvent	Temp. (°C)	Tacticity (%)			Ref.
				I	H	S	
Methyl	BuLi	DME[a]	−70	7	24	69	64)
Methyl	AmylNa	Toluene	−70	67	24	9	64)
Methyl	AmylNa	DME[a]	−70	4	23	73	64)
Methyl	OctylK	Toluene	−70	37	40	23	64)
Methyl	FluorenylK	THF	−68	7	57	37	67)
Methyl	OctylK	DME[a]	−70	12	49	39	64)
Methyl	OctylK	Pyridine	0	14	53	33	64)
Methyl	$(C_6H_{11})_3P=CH_2 \cdot LiBr$	Toluene	−60	15	39	46	63)
Methyl	$(C_6H_{11})_3P=CH_2 \cdot LiBr$	THF	−60	0	27	73	63)
Methyl	$Ph_3P=CH_2 \cdot LiBr$	Toluene	−60	79	17	4	63)
Methyl	$Ph_3P=CH_2 \cdot LiBr$	THF	−60	0	29	71	63)
Ethyl	BiphenylNa	THF	−78	6	36	58	66)
Ethyl	BiphenylK	THF	−78	15	52	33	66)
Ethyl	BiphenylK	DME[a]	−70	11	60	29	66)
Ethyl	FluorenylCs	THF	−78	5	47	48	48)
Butyl	BiphenylNa	THF	−78	7	39	52	66)
Butyl	BiphenylK	THF	−78	10	38	52	66)
Methyl	Ph_3CCaCl	DME[a]	−63	0	5	95	68)
Methyl	FluorenylCaCl	DME[a]	−63	0	2	98	68)
Methyl	DiindenylCa	DME[a]	−63	0	11	89	68)
Methyl	DiindenylCa	C_6H_6	5	43	29	28	68)
Methyl	Cp_2Ca	DME[a]	0	0	6	94	68)
Methyl	DiphenylCa	DME[a]	< 0	32	35	33	68)
Methyl	DiphenylCa	C_6H_6	0	54	24	22	68)

[a] Dimethoxyethane.

stage to syndiotactic second growth stage. This was explained by the hypothesis that there exist two separate types of active centers differing kinetically and by their stereospecific capacity.[62]

Ylid compounds $(C_6H_{11})_3P=CH_2 \cdot LiBr$ and $Ph_3P=CH_2 \cdot LiBr$ are strange initiators which were employed by Klippert et al.[63] in the polymerization of methyl methacrylate. $Ph_3P=CH_2 \cdot LiBr$ worked in the same manner as BuLi to form an isotactic polymer in toluene and a syndiotactic one in THF, while $(C_6H_{11})_3P=CH_2 \cdot LiBr$ gave rather syndiotactic polymers in both the solvents. The low stereoregularity of the polymer prepared by $(C_6H_{11})_3P=CH_2 \cdot LiBr$ in toluene was ascribed to the steric effect of the catalyst.

By means of organoalkali metal compounds methyl methacrylate can be polymerized to the polymers with various stereoregularities[64-67] as shown in Table 5. The influence of the cation in initiator, the solvent and temperature on the tacticity of the polymer was first investigated by Braun et al. using butyllithium, amylsodium and octylpotassium as initiator.[64] In a nonpolar solvent mainly isotactic polymers were obtained, whereby, the isotacticity of the polymer produced by the lithium compound was the highest and it diminished with the sodium and potassium compounds in this order. In polar solvent the syndiotacticity of the polymer was usually high when it was prepared by the lithium compound, but it decreased in the order: $Li^+ > Na^+ > K^+$. The polymerization by the potassium initiator under certain conditions gave the polymer rich in heterotacticity as described later. With solvent having higher dielectric constant such as hexamethylphosphoric triamide the microstructure of the polymer was not affected by the nature of counter ions and rather syndiotactic.[67]

Polymerization of methyl methacrylate by metallic sodium or potassium in benzene usually gives an atactic polymer. When a small amount of macrocyclic polyether or crown ether was added to this polymerization, the initiator metal was dissolved and the polymer became syndiotactic.[69, 70] Similar dissolution of insoluble initiator and/or increase in syndiotacticity were observed in the polymerization by alkali metal-naphthalene[71] and by fluorenyllithium.[72]

The stereoregularity of poly(methyl methacrylate) obtained in THF by fluorenyl- or cumylcaesium was found to be described not by Bernoullian statistics, but by first order Markov statistics.[48, 73]

Alkali metal alkoxide can initiate only the oligomerization of methyl methacrylate in the presence of methanol.[74-76] However, the activity in anionic initiation was remarkably increased in the presence of aprotic polar solvent such as hexamethyl phosphoric triamide (HMPA), dimethyl sulfoxide (DMSO) or dimethylformamide (DMF) to give high molecular poly(methyl methacrylate).[77, 78] The activity enhancement was in the order of HMPA > DMSO > DMF. The polymer obtained was usually rich in heterotacticity, especially in the case of potassium alkoxide.

Tritylcalcium chloride or fluorenylcalcium chloride gave extremely highly syndiotactic poly(methyl methacrylate) in dimethoxyethane in the temperature range of $0 \sim -63\,°C$.[68]

7 Stereospecific Polymerization of Methacrylates with Magnesium Compound

A wide variety of magnesium compounds have been used as initiators for the polymerization of methyl methacrylate as shown in Table 6. In the polymerization in toluene alkylmagnesium halides generally produce isotactic polymers whose isotacticity increases and syndiotacticity decreases as the temperature is raised from −78 to 30 °C. Among them the magnesium bromides of following alkyl groups give a 100 % isotactic poly(methyl methacrylate): isobutyl and cyclohexyl above −78 °C, octyl above −65 °C, hexyl above −30 °C and phenyl above 0 °C. The corresponding iodides seem to form the isotactic polymer at lower temperatures, while the chlorides form it less easily. In the other cases, the polymers, especially those produced at low temperature, often become stereoblock type.

The 1,4-addition product of Grignard reagent to an unsaturated ketone such as benzalacetophenone was reported to be a very effective catalyst for the preparation of highly isotactic and high molecular weight polymer.[84, 86, 87] Bis(divinylenimino)-magnesium is also an effective initiator for the preparation of isotactic poly-(methyl methacrylate) at −30 °C. Very interestingly, however, ethylmagnesium pentamethylenimide and bis(pentamethylenimino)magnesium form a highly syndiotactic polymer.[85]

The dependence of the stereoregularity of polymer upon the structure of alkyl group in initiator is much complicated in the Grignard reagent polymerization of methyl methacrylate.[79, 88] Furthermore, low reproducibility is observed also in this polymerization. Ando et al.[82] reported that the stereoregularity of polymer depended on the maximum temperature of the reaction mixture, which was resulted from the heat of polymerization, rather than the setting temperature, and that the mechanism of stereospecific polymerization by C_6H_5MgBr was different at the temperatures higher and lower than 266 °K. They postulated that the difference is attributable to the difference in the forms of reactive species. It was assumed that the polymerization by Grignard reagent was not purely anionic but proceeds through the coordination of monomer to the active centers.[88] The fact that the stereospecificity of polymerization varies largely by the alkyl group of Grignard reagent indicates that the alkyl group functions not only as an initiator anion but also a part of counter ion, which should be, therefore, in a complex form such as a dimer of the magnesium compound. Okamoto et al.[83] investigated the polymerization of methyl methacrylate using a variety of ethylmagnesium alkoxides. The alkoxide of normal alcohol did not show a clear stereospecific tendency, but the alkoxides prepared from 2-monosubstituted primary alcohol, such as isobutyl, 2-methylbutyl, and cyclohexanemethyl alcohols worked as isotactic catalysts, while secondary and tertiary alkoxides formed syndiotactic polymers. Ethylmagnesium alkoxide may exist in associated form in a nonpolar solvent (see page 15).

The nature of the above equilibria and the predominant species among three dimeric forms A, B and C will vary with the group R, solvent, temperature and concentration. It was suggested from the NMR spectra of the catalysts that the stereospecificity of polymerization depends not only on the structure of the R group but also on the form of the dimeric species.

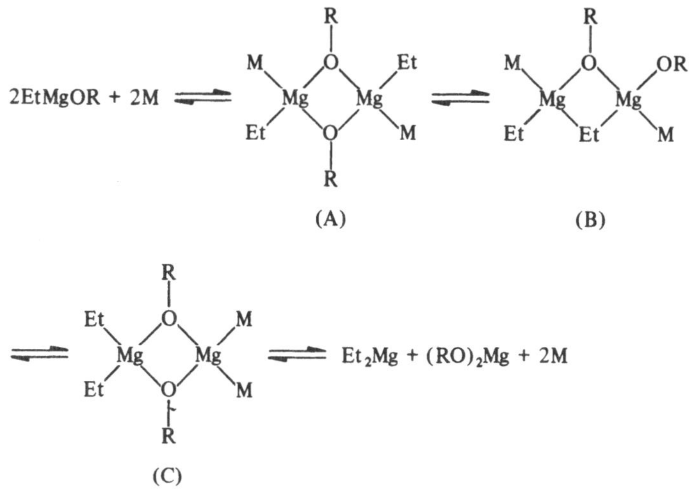

$$2EtMgOR + 2M \rightleftharpoons \text{(A)} \rightleftharpoons \text{(B)}$$

(A) (B)

$$\rightleftharpoons \text{(C)} \rightleftharpoons Et_2Mg + (RO)_2Mg + 2M$$

(C)

M: monomer or Et_2O.

In Table 7 are listed the results of the stereospecific polymerization of various methacrylates by phenylmagnesium bromide. Most methacrylates formed isotactic polymers by phenylmagnesium bromide above 0 °C in toluene. A correlation between the isotacticity of some of the polymers and the steric factor (Es) of ester groups in the monomers was reported by Matsuzaki et al.[37] Chlandra and Donaruma[89] reported that little temperature effect was found on the stereospecific polymerization of alkyl methacrylates between 0 °C and −78 °C. tert-Butyl, phenyl, and 2,3-epoxypropyl methacrylates formed polymers of low stereoregularity by this initiator. It is interesting that diphenylmethyl methacrylate gave an atactic polymer,[32] while diphenylethyl methacrylate produced a highly isotactic one by phenylmagnesium bromide.[42] By BuLi, a highly isotactic polymer was obtained from the former monomer[32] and an atactic one from the latter.[42] The stereospecificity of the polymerization changes largely not only with the size of counter ion but also with the bulkiness of monomer.

In Table 8 are summarized the microstructures of various polymethacrylates prepared by magnesium compounds other than PhMgBr. By isoBuMgBr, 2-methoxyethyl, 3-methoxypropyl, 4-methoxybutyl and 2,3-epithiopropyl methacrylates formed extremely highly isotactic polymers in toluene, and cyclopropanemethyl, 2-methyl thioethyl and 2,3-epithiopropyl methacrylates produced highly syndiotactic polymers in THF. These are in contrast with the corresponding polymers prepared by BuLi.

8 Stereospecific Polymerization with Compounds of Aluminum and other Metals

The microstructure was investigated on a number of polymethacrylates prepared by various catalysts containing aluminum and some other metals as shown in Table 9.

Table 6. Stereoregularity of poly(methyl methacrylate) prepared by magnesium compounds as initiator

Magnesium compounds	Solvent	Temp. (°C)	Tacticity (%)			Ref.
			I	H	S	
EtMgCl	Toluene	−78	30	11	59	79, 80)
EtMgCl	Toluene	−20	35	18	47	79, 80)
EtMgI	Toluene	−60	39	19	42	79, 80)
EtMgI	Toluene	>−40	100	0	0	79, 80)
BuMgCl	Toluene	−78	54	13	34	79, 80)
BuMgCl	Toluene	−30	77	11	13	79, 80)
BuMgI	Toluene	−60	25	15	59	79, 80)
BuMgI	Toluene	>−30	100	0	0	79, 80)
IsoBuMgBr	Toluene	>−78	100	0	0	79, 80)
Sec-BuMgCl	Toluene	−78	72	9	20	79, 80)
Sec-BuMgCl	Toluene	>−40	100	0	0	79, 80)
Tert-BuMgCl	Toluene	−78	99	1	0	79, 80)
HexylMgCl	Toluene	>−30	100	0	0	79, 80)
HexylMgBr	Toluene	>−30	100	0	0	79, 80)
HeptylMgCl	Toluene	−78	57	6	37	79, 80)
OctylMgCl	Toluene	−78	15	23	62	79, 80)
OctylMgCl	Toluene	−40	47	18	36	79, 80)
OctylMgBr	Toluene	>−65	100	0	0	79, 80)
CyclohexylMgCl	Toluene	>−78	100	0	0	79, 80)
CyclohexylMgBr	Toluene	>−78	100	0	0	79, 80)
PhMgBr	Toluene	−78	32	27	41	48)
PhMgBr	Toluene	> 0	100	0	0	79, 80)
PhMgBr	Et₂O	0	94	4	2	81)
PhMgI	Toluene	>−40	100	0	0	80)
Et₂Mg	Toluene	−78	17	30	54	81)
Et₂Mg	Toluene	23	25	30	46	81)
Ph₂Mg	Toluene	−78	21	31	48	82)
Ph₂Mg	Toluene	20	27	36	38	82)

	Solvent	Temp.				Ref.
EtMgOEt	Toluene	−78	26	21	53	83)
EtMgO-isoBu	Toluene	−78	91	4	5	83)
EtMgO-tert-Bu	Toluene	−78	12	14	74	83)
EtMgO-trityl	Toluene	−78	5	24	71	83)
EtMgBr-benzalacetophenone	Heptane	30	100	0	0	84)
(Et₂N)₂Mg	Toluene	−78	9	23	68	85)
(Ph₂N)₂Mg	Toluene	−78	45	30	24	85)
(⬡N)₂Mg	Toluene	−78	0	7	93	85)
(⬡N)₂Mg	THF	−78	0	8	92	85)
(⬡N)₂Mg	Et₂O	−78	5	15	80	85)
(⬡N)₂Mg	Anisole	0	53	16	32	85)
⬡N–MgEt	Toluene	−78	0	7	93	85)
⬡N–MgBr	Toluene	−78,0	100	0	0	85)
(⬠N)₂Mg	Toluene	−78∿0ᵃ	100	0	0	85)
⬠N–MgEt	Toluene	−78	100	0	0	85)

Table 6 (continued)

Magnesium compounds	Solvent	Temp. (°C)	Tacticity (%)			Ref.
			I	H	S	
Bu$_2$Mg	THF + Toluene	-50	30[b]	33[b]	37[b]	121)
			19[c]	31[c]	50[c]	121)
BuMgBr	THF + Toluene	-50	30[b]	33[b]	37[b]	121)
			18[c]	32[c]	50[c]	121)
BuMgBr + HMPA	THF + Toluene	-50	11[c]	26[c]	83[c]	121)

[a] Maximum rate at −30 °C.
[b] Methanol-soluble fraction.
[c] Methanol-insoluble fraction.

Table 7. Stereoregularity of polymethacrylates prepared by phenylmagnesium bromide in toluene

Methacrylate	Temp. (°C)	Tacticity (%)			Ref.
		I	H	S	
Ethyl	−78	94	4	2	48)
Propyl	0	88	9	3	89)
Isopropyl	0	83	12	6	89)
Isopropyl	−70	52	15	33	38)
Butyl	0	83	6	11	89)
Isobutyl	0	81	9	10	89)
Sec-butyl	0	78	13	9	89)
Tert-butyl	0	41	32	27	38)
Tert-butyl	−70	12	40	48	38)
Amyl	0	79	10	11	89)
3-Pentyl	18	73	20	7	37)
Neopentyl	−70	87	8	5	38)
Hexyl	0	81	9	10	89)
Octyl	0	80	11	9	89)
4-(2,6-Dimethyl)hexyl	18	63	30	7	37)
Decyl	0	75	15	10	89)
Lauryl	0	72	13	15	89)
Octadecyl	0	69	15	15	89)
Cyclohexyl	− 78 ∼ 18	100	0	0	37, 41)
4-Methylcyclohexyl	18	91	7	2	37, 41)
3-Methylcyclohexyl	18	81	16	3	37, 41)
2-Methylcyclohexyl	18	72	22	6	37, 41)
dl-Menthyl	20	53	27	20	41)
1-Menthyl	20	56	28	16	41)
Phenyl	18	33	39	28	37)
Phenyl	−78	20	49	31	37)
Benzyl	30	85	11	4	43)
p-Phenylbenzyl	30	82	12	6	43)
(RS)-α-Methylbenzyl	18	75	21	4	37)
Diphenylmethyl	20	28	36	36	32)
α,α-Dimethylbenzyl	30	74	16	10	34)
α,α-Dimethylbenzyl	−78	92	5	3	34)
1,1-Diphenylethyl	0	97	3	0	42)
1,1-Diphenylethyl	−78	88	9	3	42)
Trityl	30	−	−	−	32)
2-Phenylethyl	18	89	9	2	37)
2-Phenylethyl	−78	83	11	6	37)
2,3-Epoxypropyl	−78	25	40	35	53)

Table 8. Stereoregularity of polymethacrylates prepared by magnesium compounds in toluene

Methacrylate	Initiator	Temp. (°C)	Tacticity (%) I	H	S	Ref.
Ethyl	BuMgBr	−70	81	7	12	38)
Ethyl	EtMgO-isoBu	−78	17[a]	25[a]	58[a]	83)
			72[b]	14[b]	14[b]	83)
Isopropyl	BuMgBr	−70	46	13	41	38)
Tert-butyl	BuMgBr	−70	32	41	27	38)
Tert-butyl	EtMgO-isoBu	−78	5	44	51	83)
Cyclopropanemethyl	IsoBuMgBr	−78	96	2	2	53)
Cyclopropanemethyl	CyclohexylMgBr	−78	94	3	3	53)
Benzyl	BuMgCl	0	67	19	14	37)
(RS)-α-Methylbenzyl	EtMgO-isoBu	−78	64	16	20	83)
(RS)-α-Methylbenzyl	EtMgO-tert-Bu	−78	4	24	72	83)
α,α-Dimethylbenzyl	IsoBuMgBr	0	72	19	9	34)
α,α-Dimethylbenzyl	CyclohexylMgBr	0	86	9	5	34)
α,α-Dimethylbenzyl	CyclohexylMgBr	−78	33	22	45	34)
α,α-Dimethylbenzyl	Et$_2$NMgBr	0	11	38	51	34)
α,α-Dimethylbenzyl	Ph$_2$NMgBr	0	91	8	1	34)
α,α-Dimethylbenzyl	EtMgO-isoBu	−78	29	21	50	83)
α,α-Dimethylbenzyl	EtMgO-tert-Bu	−78	3	19	78	83)
1,1-Diphenylethyl	IsoBuMgBr	0	92	6	1	83)
1,1-Diphenylethyl	CyclohexylMgBr	0	91	8	1	83)
2-Phenylethyl	BuMgCl	0	81	11	8	37)
Phenyl	BuMgCl	18	18	43	39	37)
Phenyl	BuMgCl	−78	9	30	61	37)
Phenyl	BuMgBr	18	20	47	33	37)
Phenyl	BuMgBr	−40	19	69	12	37)
Phenyl	IsoBuMgBr	−78	14	74	12	37)
Phenyl	Sec-BuMgBr	−40	29	59	12	37)
2-Methoxyethyl	IsoBuMgBr	−78	80	8	12	50)
3-Methoxypropyl	IsoBuMgBr	−78	100	0	0	50)
4-Methoxybutyl	IsoBuMgBr	−78	100	0	0	50)
2,3-Epoxypropyl	IsoBuMgBr	−78	26	44	30	53)
2,3-Epoxypropyl	CyclohexylMgBr	−78	27	42	31	53)
2,3-Epithiopropyl	IsoBuMgBr	−78	95	3	2	53)
2,3-Epithiopropyl	CyclohexylMgBr	−78	100	0	0	53)
Tetrahydrofurfuryl	IsoBuMgBr	−78	7	42	51	50)
Furfuryl	IsoBuMgBr	−78	97	3		50)
N,N-Dimethylaminoethyl	IsoBuMgBr	−78	4	15	81	50)
2-Pyridylmethyl	IsoBuMgBr	−78	4	43	53	50)
2-Methylthioethyl	IsoBuMgBr	−78	98	2		50)
Cyclopropanemethyl[c]	IsoBuMgBr	−78	2	10	88	53)
Cyclopropanemethyl[c]	CyclohexylMgBr	−78	0	9	91	53)
2-Methylthioethyl[c]	IsoBuMgBr	−78	5	3	92	50)
2,3-Epoxypropyl[c]	IsoBuMgBr	−78	4	42	54	53)
2,3-Epithiopropyl[c]	IsoBuMgBr	−78	0	7	93	53)
2,3-Epithiopropyl[c]	CyclohexylMgBr	−78	0	7	93	53)
Tetrahydrofurfuryl[c]	IsoBuMgBr	−78	2	15	83	50)

[a] Methanol-insoluble fraction. [b] Methanol-soluble fraction. [c] Polymerization in tetrahydrofuran.

Table 9. Microstructure of polymethacrylates obtained by aluminum and zinc compounds

Methacrylate	Initiator	Solvent	Temp. (°C)	Tacticity (%)			Ref.
				I	H	S	
Methyl	LiAlH$_4$	Et$_2$O	−70	100	0	0	38)
Ethyl	LiAlH$_4$	Et$_2$O	−70	98		2	38)
Isopropyl	LiAlH$_4$	Et$_2$O	−70	97	1	2	38)
Butyl	LiAlH$_4$	Toluene	−70	93	5	2	38)
Sec-butyl	LiAlH$_4$	Toluene	−70	86	4	10	38)
Tert-butyl	LiAlH$_4$	Toluene	−70	90	6	4	38)
Tert-butyl	LiAlH$_4$	Et$_2$O	−70	73	9	18	38)
Cyclopropanemethyl	LiAlH$_4$	Toluene	−78	94	3	3	53)
Cyclopentanemethyl	LiAlH$_4$	Toluene	−78	100	0	0	50)
1-Menthyl	LiAlH$_4$	Et$_2$O	20	69	23	8	41)
dl-Menthyl	LiAlH$_4$	Et$_2$O	20	61	26	12	41)
D-Bornyl	LiAlH$_4$	Toluene	−70	79	16	5	38)
2,4,6-Triphenylbenzyl	LiAlH$_4$	Toluene	−70	53	24	23	38)
2-Methoxyethyl	LiAlH$_4$	Toluene	−78	39	33	38	50)
2-Methylthioethyl	LiAlH$_4$	Toluene	−78	97	3	0	50)
3-Methoxypropyl	LiAlH$_4$	Toluene	−78	87	9	4	50)
4-Methoxybutyl	LiAlH$_4$	Toluene	−78	90	10	0	50)
2,3-Epoxypropyl	LiAlH$_4$	Toluene	−78	80	15	5	50)
2,3-Epithiopropyl	LiAlH$_4$	Toluene	−78	100	0	0	53)
Tetrahydrofurfuryl	LiAlH$_4$	Toluene	−78	72	15	13	50)
Furfuryl	LiAlH$_4$	Toluene	−78	100	0	0	50)
N,N-Dimethylaminoethyl	LiAlH$_4$	Toluene	−78	81	9	10	50)
2-Pyridylmethyl	LiAlH$_4$	Toluene	−78	7	52	41	50)
Methyl	LiAlH$_4$	THF	−70	7	16	77	38)
Ethyl	LiAlH$_4$	THF	−70	9	19	72	38)
Isopropyl	LiAlH$_4$	THF	−70	5	38	57	38)
Tert-Butyl	LiAlH$_4$	THF	−70	5	51	44	38)

Table 9 (continued)

Methacrylate	Initiator	Solvent	Temp. (°C)	Tacticity (%)			Ref.
				I	H	S	
Cyclopropanemethyl	LiAlH$_4$	THF	-78	0	47	53	53)
2,4,6-triphenylbenzyl	LiAlH$_4$	THF	-70	17	47	36	38)
2-Methoxyethyl	LiAlH$_4$	THF	-78	9	13	78	50)
2-Methylthioethyl	LiAlH$_4$	THF	-78	3	11	86	50)
2,3-Epoxypropyl	LiAlH$_4$	THF	-78	3	44	53	50)
2,3-Epithiopropyl	LiAlH$_4$	THF	-78	0	13	87	53)
Tetrahydrofurfuryl	LiAlH$_4$	THF	-78	1	18	81	50)
Furfuryl	LiAlH$_4$	THF	-78	0	10	90	50)
N,N-Dimethylaminoethyl	LiAlH$_4$	THF	-78	2	2	96	50)
Methyl	Et$_2$AlNPh$_2$	Toluene	-78	1	14	85	90, 91)
p-Phenylbenzyl	Et$_2$AlNPh$_2$	Toluene	-78	2	22	76	43)
(R)-α-Methylbenzyl	Et$_2$AlNPh$_2$	Toluene	-78	4	17	79	43)
(RS)-α-Methylbenzyl	Et$_2$AlNPh$_2$	Toluene	-78	4	20	76	43)
Diphenylmethyl	Et$_2$AlNPh$_2$	Toluene	-40	8	27	65	32)
α,α-Dimethylbenzyl	Et$_2$AlNPh$_2$	Toluene	-78	1	17	82	34)
1,1-Diphenylethyl	Et$_2$AlNPh$_2$	Toluene	-40	12	34	54	42)
Methyl	SrZnEt$_4$	Toluene	-70	66	23	11	38)
Ethyl	SrZnEt$_4$	Toluene	-70	78	18	4	38)
Isopropyl	SrZnEt$_4$	Toluene	-70	65	29	6	38)
Tert-butyl	SrZnEt$_4$	Toluene	-70	68	20	12	38)
2-Methoxyethyl	SrZnEt$_4$	Toluene	-78	78	14	8	50)
2,3-Epoxypropyl	SrZnEt$_4$	Toluene	-78	50	41	9	50)
2,3-Epithiopropyl	SrZnEt$_4$	Toluene	-78	84	16	0	50)
Methyl	Et$_3$Al-TiCl$_4$(5:1)	Toluene	-78	0	6	94	92, 93)
Methyl	(IsoBu)$_3$Al-TiCl$_4$(5:1)	Toluene	-78	0	13	87	92, 93)
Methyl	BuLi-TiCl$_4$(4.5:1)	Toluene	-78	74	20	6	92, 93)
Methyl	(Ti-Allyl)$_3$Cr · Pyridine	Toluene	-15~25	5	40	55	94)

The polymerization of alkyl methacrylates initiated by $LiAlH_4$ and $SrZnEt_4$ was studied by Tsuruta et al.[38] and Iwakura et al.[50, 53] in polar and nonpolar solvents. Methacrylates gave isotactic polymers by $LiAlH_4$ in ether as well as in toluene even the monomers having polar ester groups, except for 2-methoxyethyl and 2-pyridyl-methyl esters. The isotacticity decreased with an increase in the bulkiness of the ester group. It is noticeable that the poly(methyl methacrylate) obtained in ether was 100 % isotactic. Fully isotactic polymers were also formed from cyclopentyl-methyl, 2,3-epithiopropyl and furfuryl methacrylates. The polymerization with $LiAlH_4$ in THF gave syndiotactic polymers at $-78\,^{\circ}C$, and the syndiotacticity seemed to decrease with increasing bulkiness of the monomer. Higher syndiotacticity was observed in the polymers containing polar ester groups, especially in poly(N,N-dimethylaminoethyl methacrylate). $SrZnEt_4$ usually produced isotactic polymer in toluene at $-78\,^{\circ}C$.

Dialkylaluminum diphenylamide exclusively formed syndiotactic polymers of methacrylates.[32, 43, 90, 91, 95] The polymerization by this compound at low temperature is initiated by Ph_2N^- anion and proceeds with anionic mechanism. The dimeric form of the aluminum amide at the growing chain end was postulated to account for the stereoregulation of the polymerization.[91] $C_2H_5Al(NPh_2)_2$ also gave highly syndiotactic poly(methyl methacrylate),[91] but $Al(NPh_2)_3$ formed an atactic polymer.[96]

Triethylaluminum initiates the polymerization of methyl methacrylate in the presence of donor compound such as sparteine, 2,2-bipyridyl, triethylamine and triphenylphosphine. The microstructure of the polymer was rather syndiotactic and consistent with Bernoullian statistics.[97] The polymerization by dibenzylzinc and dicyclopentadienylzinc in the presence of hexamethylphosphoric triamide gave similar results in toluene.[98]

Contrary to BuLi-TiCl$_4$, Ziegler type catalysts of $AlEt_3$-TiCl$_4$ and $AlEt_3$-Ti-(OisoPr)$_4$[92, 93, 99, 100] gave poly(methyl methacrylate) of high syndiotacticity, which is favorably compared with that of the polymer prepared by bis(pentameth-ylenimino)magnesium.[85] In the polymerization with $AlEt_3$-VOCl$_3$ catalyst stereo-block poly(methyl methacrylate) was obtained while the syndiotacticity increased upon the addition of a small amount of polar compound such as ether or amine.[101]

9 Stereospecific Polymerization of α-Substituted Acrylates other than Methacrylates

Only a few reports were published on the polymerization of α-substituted acrylates other than methacrylate.[84, 86, 87, 95, 102∿111] In Tables 10 and 11 is shown the stereoregularity of polymers obtained from methyl esters of α-ethyl-,[104, 105] α-n-propyl-,[104] α-phenyl-,[95] and α-chloroacrylic acid[84, 86] under various reaction conditions.

α-Ethylacrylate and α-n-propylacrylate gave no polymer above $30\,^{\circ}C$, because of the low ceiling temperatures. By BuLi both the monomers formed isotactic polymers in toluene. However, the stereospecificity of the polymerization is largely affected

Table 10. Stereoregularity of poly(methyl α-substituted acrylate)

α-Substituent	Initiator	Solvent	Temp. (°C)	Tacticity (%)			Ref.
				I	H	S	
Ethyl	BuLi	Toluene	−78	45	13	42	105)
Ethyl	BuLi	Toluene	0	98	2	0	105)
Ethyl	BuLi	THF	−78	8	21	71	105)
Ethyl	IsoBu₂AlNPh₂	Toluene	−40	14	12	74	105)
Propyl	BuLi	Toluene	−78	62	36	2	104)
Propyl	BuLi	Toluene	0	95	4	1	104)
Phenyl	BuLi	Toluene	−78	39	33	28	95)
Phenyl	BuLi	Toluene	−30	22	52	26	95)
Phenyl	BuLi	THF	−78	26	50	24	95)
Phenyl	OctylK	THF	−78	3	66	31	138)
Phenyl	PhMgBr	Toluene	30	33	42	25	95)
Phenyl	IsoBu₂AlNPh₂	Toluene	−78	20	32	48	95)
p-Chlorophenyl	BuLi	Toluene	−78	25	40	35	135)
p-Chlorophenyl	BuLi	THF	−78	19	61	20	135)
p-Bromophenyl	BuLi	Toluene	−40	26	55	19	135)
p-Bromophenyl	BuLi	THF	−78	21	59	20	135)

Table 11. Stereoregularity of poly(alkyl α-chloroacrylate)s obtained in toluene by various types of initiators. Ref.[86]

Alkyl	Initiator	Temp. (°C)	Tacticity (%)			M_n[a]
			I	H	S	
CH_3	Tert-BuLi	−78	12	53	35	4,300
CH_3	Breslow-Kutner catalyst[b]	30	29	38	33	32,000
CH_3	BPO	70	11	34	55	270,000
CH_3	Benzoin-UV	−50	7	33	66	109,000
C_2H_5	BuLi	0	31	44	25	4,500
C_2H_5	PhMgBr	0	27	35	38	4,500
C_2H_5	Breslow-Kutner catalyst[b]	30	65	13	23	309,000
C_2H_5	BPO[c]	70	17	35	48	362,000
C_2H_5	Benzoin-UV[c]	−50	16	30	54	181,000
$IsoC_3H_7$	BuLi	−78	53	29	18	8,000
$IsoC_3H_7$	PhMgBr	−78	–	–	–	4,900
$IsoC_3H_7$	Breslow-Kutner catalyst[b]	30	51	25	24	50,000
$IsoC_3H_7$	BPO	60	15	44	40	77,300
$IsoC_3H_7$	Benzoin-UV	−65	8	18	74	2,100

[a] Number average molecular weight by either vapor pressure depression for $\overline{M}_n < 50,000$ or osmotic pressure for $\overline{M}_n > 50,000$.
[b] Polymerization in heptane-ether mixture. See Ref.[84] for the preparation of the catalyst.
[c] Polymerization in heptane.

by temperature. The triad isotacticity was only 45% in the poly-α-ethylacrylate produced at −78 °C but reached 98% in the polymer formed at 0 °C, while the tacticity of poly(methyl methacrylate) increases only slightly with increasing polymerization temperature. The poly-α-ethylacrylate prepared at low temperature was a mixture of isotactic and syndiotactic polymers, as described later.

Methyl α-phenylacrylate formed an isotactic polymer by BuLi at −78 °C in toluene, although the tacticity was not so high. By increasing the polymerization temperature the isotacticity decreased in association with the increases in the heterotactic and syndiotactic contents. The proportion I : H : S of the polymer obtained above −45 °C was close to 1 : 2 : 1, suggesting that the stereoregulation in the propagation step was random. In a polar solvent, such as THF and dimethoxyethane the sterically random polymer was formed even at −78 °C. No polymer was obtained at 0 °C whereas the radical polymerization in bulk gave a trace of polymer at 60 °C whose tacticity was random. Substitution by chlorine or bromine at the para position of the phenyl group caused to increase the heterotactic contents, which will be discussed later. Poly(methyl α-phenylacrylate) prepared by $CaZnEt_4$ or $SrZnEt_4$ was fractionated by methyl ethyl ketone, where the insoluble fraction showed some crystallinity although the tacticity of the polymer was not determined.[102]

Alkyl α-chloroacrylates gave syndiotactic polymers of high molecular weight by radical initiator. However, the polymerization of these monomers with various anionic initiators including phenylmagnesium bromide formed only low molecular weight materials. Breslow-Kutner catalyst was very effective for the synthesis of the highly isotactic, high molecular weight crystalline polymer of the α-chloroacrylate[84, 86] (Table 11).

10 Multiplicity of Active Species in the Anionic Polymerization of α-Substituted Acrylates

Poly(methyl methacrylate) which has a low heterotactic content compared with isotactic and syndiotactic one was first called a "stereoblock" polymer,[1, 112] but afterwards it was found that the polymer often was a mixture of isotactic and syndiotactic polymers.[113] The two tactic isomers in the mixture could be recognized by thin layer chromatography on silica-gel,[114, 115] and recently separated by a competitive adsorption method.[116]

The simultaneous formation of isotactic and syndiotactic polymers indicates that there are at least two types of active species which are not in dynamic equilibrium in the polymerization mixture. This was more clearly demonstrated by the polymerization of ethyl methacrylate by BuLi in toluene.[11, 117, 118] It formed highly isotactic polymers above −20 °C, but gave less stereoregular polymers at lower temperatures. The latter polymers consisted of two fractions as shown in Table 12: one was isotactic and soluble in methanol[1], and the other was syndiotactic and insoluble in methanol. The yield of the insoluble fraction decreased steeply with the elevation of polymerization temperature and became zero at −20 °C, while the yield of the soluble fraction remained almost constant. Consequently, the number of syndiotactic active species decreased at higher temperature, while that of isotactic one was constant. The syndiotactic species may be thermally unstable, but much more reactive than the isotactic one. When a small amount of methanol, water or chloroform was added to the polymerization mixture, the syndiotactic species reacted preferentially with the polar compound because of its higher reactivity, resulting in the preferential deterioration of syndiotactic site and, consequently, the decreased yield accompanied with the increased isotacticity of the total polymer. The poly-(methyl α-ethylacrylate), which was prepared in toluene by BuLi at −78 °C, could be similarly fractionated into isotactic and syndiotactic fractions by benzene-cyclohexane mixture.[105]

The coexistence of two types of active species, which were not in dynamic equilibrium, was also postulated in the polymerizations of methyl methacrylate by the sodium salt of oligomeric α-methylstyrene in THF,[119] by AlEt$_3$-V(NEt$_2$)$_4$ catalyst[120] and by BuMgBr-Bu$_2$Mg system in THF-toluene mixture,[121] and in the polymerization of 2-methoxyethyl methacrylate by lithium tert-butoxide in benzene.[62] The concept of the multiple active species described above means that

---

1 In the polymerization of methyl methacrylate initiated by organometallic compound the formation of low molecular weight polymer is always observed. However, polymers of ethyl, propyl, and butyl methacrylate often contain methanol-soluble fraction, even if the fraction is not so low molecular weight. Therefore, it is necessary to recover the soluble polymer, when methanol is used as precipitant for the above polymers.

Table 12. Polymerization of ethyl methacrylate in toluene by BuLi at various temperatures for 24 hr[a]. Ref.[118]

Temp. (°C)	Insoluble in MeOH					Soluble in MeOH				
	Yield (%)	$M_n \times 10^{-3}$	Tacticity (%)			Yield (%)	$M_n \times 10^{-3}$	Tacticity (%)		
			I	H	S			I	H	S
−78	18	28.9	19	33	48	81	6.64	80	14	6
−40	6	40.5	44	25	31	88	7.36	79	14	7
−20	0	−	−	−	−	94	8.25	81	13	6
0	0	−	−	−	−	94	7.67	78	17	5
30	0	−	−	−	−	60	4.29	70	21	9

[a]   Ethyl methacrylate 10 mmol, Toluene 10 ml, BuLi 0.3 mmol.

the stereospecificity of the polymerization is dominated by the relative amounts of the isotactic and syndiotactic species. This seems to be applicable in many cases of stereospecific polymerization of α-substituted acrylates, although it may be not always so. In this case the polymer becomes more or less heterogeneous with respect to the tacticity.

The complete illustration is not possible at present on the formation of the two active species. Each of the chain ends enabling the isotactic and syndiotactic propagation may possess respective conformation which is rigid enough to keep the mode of addition of incoming monomer in the same tacticity. There are some possibilities that the polymerization proceeds through the formation of stereocomplex, especially in the case of methyl methacrylate. It has been demonstrated that the formation of stereocomplex between isotactic and syndiotactic polymers plays an important role in the stereospecific template polymerization of methyl methacrylate by radical initiator[122–127] and by BuMgCl.[128] The stereocomplex of poly(methyl methacrylate) was reported to be very stable in a certain solvent.[129–131] However, the complex formation was not observed in the cases of the polymers of ethyl, isopropyl and tert-butyl methacrylates.[132] The possibility of the formation of stereoblock polymer also cannot be eliminated. A possible model of stereoblock formation has been proposed by Coleman and Fox, in which two different active centers are in a dynamic equilibrium, both adding the monomer with different tacticity.[133, 134] It has been suggested that in addition to the normal uncomplexed ion-pair form there exist the growing chain ends complexed with one or even more molecules of methoxide in the polymerization of methyl methacrylate in toluene with butyllithium[5, 13].

11 Heterotactic Polymers

A polymer having more than 50 % portion of heterotactic triad should be defined to be a heterotactic polymer, since the heterotactic content could attain at most to 50% even in a sterically random polymer. The formation of heterotactic polymer seems to be much interesting because a higher order of stereoregulation would be required for the heterotactic polymerization than that for isotactic or syndiotactic one.

As mentioned already in the polymerization of methyl α-phenylacrylate a sterically random polymer was obtained. However, the fraction of heterotactic triad in the polymer did not exceed 50 % under any polymerization conditions with BuLi initiator[95] (Table 10). On the other hand, methyl α-(p-bromophenyl)acrylate and methyl α-(p-chlorophenyl)acrylate gave the polymers whose heterotactic contents were 59 and 61 %, respectively, in THF by BuLi at −78 °C[135] (Table 10). The tacticity was determined by the ^{13}C-NMR spectra of poly(2-phenylallyl acetate)s derived from the original polymers via poly(2-phenylallyl alcohol)s. The bulkiness of the α-substituent seems to be important for the heterotactic propagation in these cases, as pointed out by Nozakura et al. in the formation of heterotactic poly(vinyl alcohol) derived from poly(vinyl trimethylsilyl ether).[136, 137] However, methyl α-phenylacrylate formed a heterotactic polymer whose heterotactic content was 66 %, by octylpotassium in THF[138] (Table 10).

In the literatures we can find out a number of data which indicate the formation of heterotactic polymers, although the authors have not commented about it. These include poly(methyl methacrylate) and poly(ethyl methacrylate) obtained by octylpotassium,[64] biphenylpotassium,[66] fluorenylpotassium (Table 5),[67] benzylcalcium[139] or potassium alkoxide[77] in polar solvent, and poly(phenyl methacrylate) prepared by n-, iso- and sec-butylmagnesium bromide in toluene at low temperature (Table 8).[37] These clearly indicate that not only the size of the substituent of monomer but also the nature of the counter cation or the polymerization medium is important factor for the formation of heterotactic polymer.

12 Slow Growth Polymerization

As was mentioned previously the insufficient mixing of initiator and monomer solutions sometimes causes the fluctuation of the data of polymerization. Slow growth polymerization[20, 21, 140, 141] is a unique method carrying out ionic polymerization at low temperature without agitation. It might have something in common with the polyphase or proliferous polymerization of vinyl ether reported by Schildknecht et al.[142] The method is as follows: An initiator solution is slowly added to a cooled monomer solution so as to rest on it as a separate liquid phase, and the polymerization is allowed to proceed without stirring. After a given polymerization time the reaction vessel is cooled in liquid nitrogen and carefully broken. Then the reaction mass is sliced into several portions, and the polymer is isolated from each portion and characterized.

The results of the methyl methacrylate polymerization by this method in toluene are shown in Table 13. The notations I, II, III,. . . , indicate the portions from top to bottom of the reaction mass. The isotacticity and the molecular weight of the polymer obtained by slow growth method were much higher than those of the polymer obtained by ordinary method. In the former method the initiation occurred only at the interface between the initiator and monomer solutions and the active

Table 13. Slow growth polymerization of methyl methacrylate in toluene by BuLi at $-78\,^{\circ}$C for 24 hr[a]. Ref.[21]

	Yield (%)	Tacticity (%)			$M_n \times 10^{-3}$	Number of polymer chain x 10^3 (mmol)	Li compd[c] (mmol)
		I	H	S			
I	19.5	84	11	5	29.3	6.7	0.50
II	9.9	80	13	7	35.2	2.8	0.01
III	23.3	80	14	6	52.1	4.5	0.02
O[b]	72	70	19	11	21.1	33.9	–

[a] Monomer 10 mmol, BuLi 0.5 mmol, toluene 10 ml.
[b] Polymerization by ordinary method.
[c] Amount of lithium compound contained in the reaction mixture after the polymerization.

Table 14. Slow growth polymerization of methyl methacrylate in tetrahydrofuran by BuLi at $0\,^{\circ}$C for 24 hr[a]. Ref.[21]

	Yield (%)	Tacticity (%)		
		I	H	S
I	1.0	68	14	18
II	3.3	55	18	27
III	0.8	54	19	27
O[b]	15.5	19	35	46

[a] Monomer 10 mmol, BuLi 0.5 mmol, tetrahydrofuran 10 ml.
[b] By ordinary method.

species thus formed grew up to high molecular chain during the diffusion from top to bottom. Most of the BuLi used remained in the initiator layer without initiating the polymerization even after the completion of the reaction. As a result the molecular weight of the polymer was much higher and increased from top to bottom. A remarkable increase in the isotacticity was also observed in the polymerization in THF as shown in Table 14. This increase was caused by the formation of stereocomplex at the interface of the initiator and monomer solutions. The complex allowed the isotactic species to diffuse preferentially, because of its higher concentration than the syndiotactic one. Slow growth polymerizations of various methacrylates by several initiators were also studied.[143]

13 Asymmetric-Selective Polymerization of Methacrylates

Asymmetric-selective (or stereoelective) polymerizations were attempted with methacrylic esters of racemic menthyl,[144] α-methylbenzyl[145, 146] and sec-

Table 15. Polymerization of α-methylbenzyl methacrylate with diethylmagnesium-chiral alcohol and cyclohexylmagnesium chloride-(−)-sparteine in toluene at −78 °C. Ref.[147]

Chiral alcohol (R*OH)	Time (hr)	Yield (%)	ηsp/C dl/g	$[\alpha]_D^{20}$ (O.P.)[a] of polymer	Tacticity (%)		
					I	H	S
(−)-cis-Myrtanol	2.5	19	0.36	−2.7(2)	7	29	64
(−)-Borneol	2.5	11	0.20	−0.9(1)	−	−	−
(−)-Menthol	2.5	34	0.26	−1.2(1)	3	28	69
(−)-2-Methylbutanol	3.0	6	−	∼0	−	−	−
(+)-1-p-Menthen-9-ol	4.0	5	−	∼0	−	−	−
Quinine	48	11	0.38	+14.9(12)	23	26	51
Cinchonine	144	18	0.64	−21.3(17)	34	32	34
Cinchonidine	22	46	1.08	+1.2(1)	7	19	74
(−)-cis-Myrtanol-(−)-sparteine[b]	96	20	−	−42.9(34)	39	29	32
CyclohexylMgCl-(−)-sparteine	17	11	−	−115.9(93)	96	2	2
CyclohexylMgCl-(−)-sparteine	3.5	66	0.30	+50.9(96)[c]	90	6	4

a Optical purity (%).
b [Myrtanol]/[Sparteine] = 1/1.2.
c $[\alpha]_D^{20}$ (O.P.) of unreacted monomer.

butyl[146] alcohols using optically active catalysts. However, almost no selectivity
was observed. When ethylmagnesium alkoxide was prepared from diethylmagnesium
and alkaloid, such as quinine or cinchonine the alkoxide functioned to some extent
as an asymmetric-selective catalyst in the polymerization of (RS)-α-methylbenzyl
methacrylate, as shown in Table 15.[147] Ethylmagnesium alkoxide complexed with
(−)-sparteine showed higher selectivity in this polymerization. It was found that the
higher the selectivity was, the higher the isotacticity of the polymer became. When
Grignard reagent such as cyclohexylMgCl, cyclohexylMgBr or menthylMgCl was
complexed with (−)-sparteine, the complex polymerized (RS)-α-methylbenzyl
methacrylate with extremely high asymmetric-selectivity in toluene at −78 °C. In an
appropriate stage of the polymerization the optical purity of the polymer and that of
the unreacted monomer reached above 90 %.[147–149]

The asymmetric selectivity arises from the preferential formation of (S)-elective
center at the beginning followed by the formation of (R)-elective center after the
consumption of most of the (S)-monomer. The copolymerization of the (RS)-mono-
mer and methyl methacrylate by this complex yielded a highly isotactic copolymer
in which the (S)-monomer predominantly incorporated over the (R)-monomer. On
the other hand, in the copolymerization with α,α-dimethylbenzyl methacrylate only
the homopolymer of α-methylbenzyl methacrylate was obtained with the same as-
symmetric selectivity as in the homopolymerization of this monomer. The results
indicate that the steric interaction between the methyl group at the α-position of
benzyl ester and the (−)-sparteine moiety of the catalyst plays an important role
in the stereoelection of the polymerization.

Polymerization of sec-butyl methacrylate[148] 2,3-epoxypropyl methacrylate[83]
by the same catalyst systems also showed asymmetric selectivity to some extent. In
the polymerization of the latter monomer the sparteine complex gave a 100 % isotactic polymer, while cyclohexylMgBr alone formed only an atactic polymer.

14 Properties of Polymethacrylates in Relation to Stereoregularity

A number of physical properties of polymethacrylates have been reported to depend
on their stereoregularity. Among these properties glass transition temperature, Tg,
and NMR spin-lattice relaxation time, T_1, may be easy of access even in the laboratory of polymer synthesis. Solubility of polymer is also dependent on the stereoregularity, which has been capitalized on the fractionation of the polymer in terms
of the tacticity as described previously.

The Tg of polymethacrylates increases as the bulkiness of side chain increases
and its flexibility decreases.[150, 151] Tg also correlates with the tacticity of poly(α-
substituted acrylate) and is lower for isotactic polymer than for syndiotactic one.
Good correlation was obtained between the dyad tacticity and Tg. Therefore, the Tg
value of 100 % isotactic or 100 % Syndiotactic polymer could be determined by the
extrapolation of observed values even if it has not been prepared.[78, 86, 107, 152, 153]
The Tg for purely syndiotactic poly(methyl methacrylate) thus obtained is

160 °C.[154] Tg also depends on the molecular weight of the polymer, so the experimental value must be corrected by the relationship

$$Tg\infty = Tg(\text{experimental}) + K/\overline{M}_n$$

where the $Tg\infty$ is the glass transition temperature at infinite molecular weight and
the correction constant K is the order of 10^5.[86, 107, 154]

It should be noted that the steric configuration affects Tg only in the polymers
of 1,1-disubstituted ethylenes $(-CH_2CXY-)_n$ and, conversely, Tg is independent of
configuration when one of the substituents is replaced by hydrogen. This empirical
observation was put on a theoretical basis using the Gibbs-DiMargio theory, which
was summarized by the relationship

$$Tg(\text{syndiotactic}) - Tg(\text{isotactic}) = 0.59 \frac{\Delta\epsilon}{k}$$

where $\Delta\epsilon$ is the difference in the Gibbs-DiMargio flex energy between the syndiotactic and isotactic polymers and k is Boltzmann's constant[152]. This equation was
applied to poly-(alkyl methacrylate) and poly(alkyl α-chloroacrylate).[86, 107, 153]

NMR relaxation studies of chemically shifted nuclei give valuable information on
the local molecular motion of polymer in solution. The relaxation time, especially
the spin-lattice relaxation time, T_1, of polymer in solution can now easily be measured
by using Fourier transform NMR spectrometer. Recently several works[26-28, 155-158]
were reported on the relationship between the T_1 of polymethacrylate and its
tacticity. Most of them[27, 28, 156-158] were concerned with the ^{13}C-T_1, which
could be easily described in terms of the segmental reorientation of the polymer. The
T_1 of the carbon in isotactic polymethacrylate was found to be consistently longer
than that of the corresponding carbon in syndiotactic polymer. The correlation
times, τ's, for various poly(alkyl methacrylate)s, are shown in Table 16.[27, 28] These
values were calculated from the observed T_1 and is related to the segmental motion
of the polymer in solution. The results strongly indicate that the isotactic polymer
chain has a greater segmental mobility compared with the syndiotactic one. The
segmental mobility appears to decrease with an increase in the bulkiness of the ester
group in both isotactic and syndiotactic polymers. The α-methyl group in isotactic
polymer has also a greater freedom of internal rotation than that in syndiotactic one.
The activation energy for the internal rotation of α-methyl group is 17.1 kJ/mol for
isotactic polymer and 30.9 kJ/mol for syndiotactic one.[28]

In the case of ^1H-T_1 the mechanism of relaxation is rather complicated and the
observed T_1 cannot be directly related to the segmental mobility of polymer. However, the ^1H-T_1's of poly(alkyl methacrylate)s were found to be parallel with ^{13}C-T_1's
namely, the T_1 of the protons in isotactic polymer was always longer than that of the
comparable protons in syndiotactic polymer.[28] The study of the ^1H-T_1 in partially
deuterated and undeuterated poly(methyl methacrylate)s indicated that the T_1's of
the protons in the side chain groups were mainly controlled by the segmental motion
of the polymer chain and that the longer T_1's of side chain protons in isotactic
polymer were due to the greater mobility of the polymer chain compared with the
syndiotactic one.[29]

Table 16. ^{13}C T_1 and correlation times (τ) of poly(alkyl methacrylate)s in toluene-d$_8$ at 110 °C. Ref.[27, 28]

Alkyl	Tacticity	T_1 (msec)		$\tau \times 10^{11}$(sec)		$\tau_R \times 10^{11a}$ (sec)
		CH$_2$	α-CH$_3$	CH$_2$	α-CH$_3$	
CH$_3$	Isot.	279	494	8.35	3.14	5.0
CH$_3$	Syn.	104	220	22.4	7.06	10.3
C$_2$H$_5$	Isot.	192	419	12.13	3.71	5.3
C$_2$H$_5$	Syn.	86	219	27.1	7.09	9.6
CH(CH$_3$)$_2$	Isot.	129	347	18.1	4.47	5.9
CH(CH$_3$)$_2$	Syn.	77	175	30.2	8.87	12.6
C(CH$_3$)$_3$	Isot.	74	237	31.5	6.55	8.3
C(CH$_3$)$_3$	Syn.	36	132	64.7	11.76	14.4

a ^{13}C correlation time for internal rotation of α-methyl group.

15 Stereospecificity in the Copolymerization of Methacrylates

Otsu et al.,[159] observed in radical copolymerization that the reactivity of methacrylate toward β-chloroethyl methacrylate radical linearly depended on Taft's $\sigma*$ value of the ester alkyl group in the monomer ($\rho*$=0.13), but there was no steric effect.

The monomer reactivity ratios in the anionic copolymerization of methacrylates are listed in Tables 17 and 18.[33, 34, 42, 43, 160–164] In the copolymerizations of methyl methacrylate(M$_1$) with other methacrylates(M$_2$) by BuLi (Table 17), a roughly linear relationship was observed between log l/r$_1$, i. e. the relative reactivity of methacrylate toward methyl methacrylate anion, and the electron density on the β-carbon of M$_2$ methacrylate represented by its chemical shift, but no correlation

Table 17. Monomer reactivity ratios in the copolymerization of methyl methacrylate(M$_1$) with various methacrylates(M$_2$) by BuLi in toluene and in tetrahydrofuran at −78 °C. Ref.[43]

Ester group in M$_2$	In toluene		In tetrahydrofuran	
	r_1	r_2	r_1	r_2
Ethyl	1.10	0.38	1.13	0.52
Isopropyl	2.75	0.20	2.29	0.42
Tert-butyl	4.41	0.02	5.07	0.02
Benzyl	0.59	1.60	0.70	1.46
(RS)-α-Methylbenzyl	1.68	0.78	2.04	1.52
Diphenylmethyl	0.57	0.55	1.11	1.57
α,α-Dimethylbenzyl	10.1	0.56	2.59	2.00
1,1-Diphenylethyl	7.45	0.56	1.13	1.62
Trityl	6.28	0.13	0.62	0.62

Table 18. Monomer reactivity ratios in the anionic copolymerization of methacrylates

M_1	Ester group M_2	Temp. (°C)	Solvent	Initiator	r_1	r_2	Ref.
Methyl	Benzyl	0	THF	$C_{10}H_8Na$	0.89	1.1	163)
Methyl	Benzyl	0	Toluene	BuLi	0.70	1.84	163)
Methyl	Benzyl	0	THF	BuLi	1.02	2.04	163)
Methyl	Benzyl	0	Toluene	PhMgBr	0.95	2.87	163)
Methyl	Benzyl	0	THF	PhMgBr	1.16	2.71	163)
Methyl	Benzyl	0	Toluene	$LiAlH_4$	0.44	1.04	163)
Methyl	Benzyl	0	THF	$LiAlH_4$	0.97	1.98	163)
Methyl	Cyclohexyl	30	Dioxane	$C_{10}H_8Na$	1.4	0.8	164)
Methyl	Cyclohexyl	30	Dioxane	Li	1.57	1.27	164)
Methyl	Cyclohexyl	30	Dioxane	Na	2.09	0.82	164)
Methyl	Cyclohexyl	30	Dioxane	K	1.90	0.55	164)
Methyl	Cyclohexyl	30	THF	Li	1.27	0.84	164)
Methyl	Cyclohexyl	30	THF	Na	1.49	0.85	164)
Methyl	Cyclohexyl	30	THF	K	2.02	0.86	164)
Methyl	Cyclohexyl	30	Benzene	Li	1.34	1.84	164)
Methyl	Cyclohexyl	30	Benzene	Na	1.96	0.49	164)
Methyl	Cyclohexyl	30	Benzene	K	2.19	0.73	164)
Methyl	2-Methoxyethyl	20	Benzene	Tert-BuOLi	1.51	2.10	162)
Methyl	2-(N,N-dimethylamino)ethyl	20	Benzene	Tert-BuOLi	1.20	0.40	162)
2-Methoxyethyl	2-(N,N-dimethylamino)ethyl	20	Benzene	Tert-BuOLi	1.40	0.60	162)
(RS)-α-Methylbenzyl	Trityl	−78	Toluene	BuLi	4.30	0.03	165)
(S)-α-Methylbenzyl	Trityl	−78	Toluene	BuLi	8.55	0.005	165)
(RS)-α-Methylbenzyl	Trityl	−78	THF	BuLi	0.39	0.33	165)
(S)-α-Methylbenzyl	Trityl	−78	THF	BuLi	0.39	0.33	165)

was found between the reactivity and Taft's steric parameter of the ester group in
M_2 [42, 161]. Although the overall rate and the stereospecificity of the copolymerization in toluene were much different from those in THF, the monomer reactivity
ratios were not so much affected by the polymerization medium, except for the
cases described later.

The stereoregularity of copolymers formed from various combinations of methacrylates usually lays in between those of homopolymers formed from each of the
comonomers. However, when methyl methacrylate(M_1) was copolymerized in toluene
with a bulky monomer(M_2), such as α,α-dimethylbenzyl, 1,1-diphenylethyl or trityl
methacrylate, not only the M_2 monomer showed an extremely low reactivity
(Table 17), but also the yielded copolymer was a mixture of two or three types of
polymers.[34, 42, 161, 166] By the treatment of the copolymer with methanolic HCl,
only the tertiary ester group in the M_2 monomer unit was selectively hydrolyzed,
and the resultant poly(methyl methacrylate-co-methacrylic acid) was fractionated
into methanol soluble and insoluble portions. The insoluble portion was essentially a
homopolymer of methyl methacrylate which contained its unit more than 90 %,
while the soluble portion was a copolymer of M_1 and M_2 with lower stereoregularity
(Table 19). Furthermore, the insoluble portion resulted from the product obtained
in the copolymerization at 0 °C was proved to be a mixture of isotactic and syndiotactic polymers by thin layer chromatography. Here, again, the multipolicity of active species: isotactic and syndiotactic active sites for methyl methacrylate polymerization and an active site for copolymerization, was observed. If the copolymerization
of methacrylates proceeds stereoregularly with a specific conformation of the growing chain, the reactivity of such a chain end may be different from that of the
labile random coil chain end which gives an atactic polymer. In such a case, if a
monomer once forms a chain of the specific conformation, which may be a helix, in
the initial stage of the copolymerization, it is probable that by a steric interference
only the same monomer can add to the growing chain end to form a homopolymer,
and the other comonomer is excluded from the propagation reaction. On the other
hand, if both the monomers are copolymerized in the initial stage, the stereoregulation at the chain end will be disturbed and both monomers can continue to add to
the end, forming a less stereoregular copolymer. The extremely low reactivity of the
three bulky methacrylates in the copolymerization described above can be explained
by the difficulty of the incorporation of these monomers into the stereoregular
polymer chain of methyl methacrylate.[161]

The donor element in the ester group of 2-methoxyethyl methacrylate was
proved to affect strongly the stereospecificity in the copolymerization of this monomer with methyl methacrylate by t-BuOLi in a non-polar solvent.[162] The copolymer had higher isotacticity at very low conversion. However, the isotacticity
remarkably decreased during the first few percent of conversion, and then the copolymer became rather syndiotactic. The syndiotacticity of the copolymer obtained
at high conversion was higher than that of poly(2-methoxyethyl methacrylate) even
if the copolymerization was performed at very low 2-methoxyethyl methacrylate/
methyl methacrylate ratio and even up to practically 100 % conversion.

The copolymerization by BuLi in THF seems to give homogeneous copolymers
with respect to the sequence distribution even if tertiary benzyl methacrylates were

Table 19. Copolymerization of methyl methacrylate(M_1) with methacrylate(M_2) of tertiary alcohol by BuLi in toluene

Ester group of M_2	Temp. (°C)	Yield (%)	Insoluble fraction					The hydrolyzed copolymer — Soluble fraction					Ref.
			Yield (%)	M_1 content (mol %)	Tacticity (%) I	H	S	Yield (%)	M_1 content (mol %)	Tacticity (%) I	H	S	
α,α-Dimethyl-benzyl	−78	21	8	97	67	22	11	13	75	68	24	8	34)
	0	66	4	79	57	20	23	62	56	72	22	6	34)
1,1-Diphenyl-ethyl	−78	26	12	96	63	24	13	14	62	56	32	13	42)
	0	90	8	85	41	29	30	82	46	56	33	11	42)
Trityl	−78	24	10	100	83	11	6	14	69	56	31	13	161)
	0	27	8	92	51	23	26	19	68	50	34	16	166)

employed as the comonomer.[42] Klesper et al. reported the complete assignment of the ^{1}H NMR spectrum of methyl methacrylate-methacrylic acid copolymer prepared by radical initiator. They also demonstrated that the configuration of the copolymer can be calculated in terms of the conditional parameter P_{ij} and the configurational parameter σ_{ij}, if terminal model statistics prevails in the copolymer. The P_{ij} is the probability of finding the M_j unit as the next neighbor of the M_i unit, and σ_{ij} is the probability of generating a meso dyad when a new monomer unit M_j is formed at the M_i end of a growing chain.[167, 168] The microstructure of poly(methyl methacrylate-co-trityl methacrylate) obtained by BuLi in THF at -78 °C was analyzed, and the calculation was also applied to the data. Assuming $r_1 = r_2 = 0.62$, $\sigma_{11} = 0.25$ and $\sigma_{22} = 0.94$, $\sigma_{12} = \sigma_{21} = 0.25$ was obtained as the best fitting values with the experiment. The validity of the value was confirmed by the peak intensities of α-CH$_3$ resonances in the ^{1}H NMR spectra of poly(methyl methacrylate-co-methacrylic acid)s which were derived from the copolymers by the selective hydrolysis of trityl ester with methanolic HCl.[33]

Yamada and Tanaka reported that the best fit was attained assuming $\sigma_{12} = \sigma_{21} = \sqrt{\sigma_{11}\sigma_{22}} = 0.41$ for the radically prepared copolymers of methyl methacrylate and trityl methacrylate.[169]

Similarly the coisotactic parameters were obtained in the copolymerizations between methyl methacrylate(M_1) with other methacrylates(M_2),[35] and between α-methylbenzyl methacrylate(M_1) with trityl methacrylate(M_2)[165] by radical and BuLi initiators in THF. The parameters are shown in Table 20.

When (RS)-α-methylbenzyl methacrylate(M_2) was copolymerized with methyl methacrylate(M_1) by BuLi at -78 °C, the monomer reactivity ratios were obtained to be $r_1 = 1.68$, $r_2 = 0.78$ in toluene and $r_1 = 2.04$, $r_2 = 1.52$ in THF.[161] These values were not changed if (S)-α-methylbenzyl methacrylate was employed instead of the racemic monomer.[170] The copolymerization of (RS)-α-methylbenzyl methacrylate-(M_1) and trityl methacrylate(M_2) by BuLi in THF also gave the same r_1 and r_2

Table 20. Coisotactic parameters in the copolymerization of methyl methacrylate(M_1) and other methacrylates(M_2) by AIBN at 60 °C and by BuLi at -78 °C in tetrahydrofuran. Ref.[35]

Ester group in M_2	AIBN Copolymerization			BuLi Copolymerization		
	σ_{22}	$\sigma_{12} = \sigma_{21}$	σ_{11}	σ_{22}	$\sigma_{12} = \sigma_{21}$	σ_{11}
Diphenylethyl	0.47	0.32	0.23	0.42	0.17	0.25
Dimethylbenzyl	0.36	0.30	0.23	0.20	0.22	0.25
Tert-butyl	0.25	0.24	0.23	0.40	0.26	0.25
Diphenylmethyl	0.24	0.23	0.23	0.10	0.27	0.25
Phenyl	0.27	0.25	0.23	0.21	0.50	0.25
1-Naphthyl	0.31	0.27	0.23	0.20	0.34	0.25
Trityl	0.71[a]	0.43[a]	0.24[a]	0.94[b]	0.25[b]	0.25[b]
Trityl[c]	–	–	–	0.94	0.25	0.20

[a] Ref.[169]

[b] Ref.[33].

[c] M_1 = (RS)-α-Methylbenzyl methacrylate.

values as those obtained with the (S)-monomer within the experimental error. However, the copolymerization in toluene gave different reactivity ratios as follows: $r_1 = 8.55$ and $r_2 = 0.005$ for (S)-monomer, while $r_1 = 4.30$ and $r_2 = 0.03$ for (RS)-monomer,[165] where the reactivity of trityl methacrylate was very low in both cases.

The isotacticity is higher in poly[(R)-α-methylbenzyl methacrylate] prepared by BuLi in toluene than in the polymer of the racemic monomer.[49] However, the stereoregularity of poly[(S)-α-methylbenzyl methacrylate-co-methyl methacrylate] is mostly the same as that of poly[(RS)-α-methylbenzyl methacrylate-co-methyl methacrylate], regardless of their compositions except for low methyl methacrylate contents. This indicates that the isotactic placement of the (R)- or (S)-monomer to the growing chain ending in the antipode monomer unit is less favorable than to the anion of the same monomer unit, while the stereospecificity in the addition of α-methylbenzyl methacrylate to methyl methacrylate unit should be the same between the (R)- and (S)-monomers.[165]

On the other hand, the copolymerization of (S)-α-methylbenzyl methacrylate (M_1) and trityl methacrylate(M_2) by BuLi in toluene yielded a mixture of an isotactic polymer of M_1 and a less stereoregular copolymer, similarly to the copolymerization of methyl methacrylate(M_1) and trityl methacrylate(M_2) in this solvent. If the (RS)-monomer was used instead of (S)-isomer, only the copolymer of low stereoregularity was produced, although it had a wide distribution as to the composition. The low reactivity of trityl methacrylate in the copolymerization even with (RS)-α-methylbenzyl methacrylate may be also attributed to the steric hindrance due to the bulkiness of the trityl ester, since the homopolymerization of this monomer itself proceeded extremely slowly in toluene.[165]

When (S)-α-methylbenzyl methacrylate(M_1) and trityl methacrylate(M_2) are copolymerized at a smaller ratio of $[M_1]_0/[M_2]_0$ by BuLi in THF, the composition m_1/m_2 of the initial copolymer will be larger than $[M_1]_0/[M_2]_0$ because of the monomer reactivity ratios being $r_1 = 0.39$ and $r_2 = 0.33$,[165] and as the copolymerization proceeds more and more, the molar ratio of the remaining monomers $[M_1]/[M_2]$ becomes smaller and smaller. Therefore, the polymer chain formed in the later stage of the copolymerization will be composed of long M_2 sequences linked with an M_1 monomer unit. Since trityl methacrylate polymerizes in isotactic manner, the copolymer increases its isotactic content as the copolymerization proceeds, although the stereoregularity of the initial copolymer is low. The homopolymer of (S)-α-methylbenzyl methacrylate had a specific rotation of $[\alpha]_{589}^{20} -103.8°$, and the initial copolymers with trityl methacrylate showed also negative rotations whose absolute values were slightly higher than that calculated from the composition. However, the copolymers obtained in high yield had positive rotations of abnormally large values in spite of the small contents of the chiral ester group (Table 21). This abnormal chiroptical property of the copolymers was elucidated by the helical structure of isotactic sequences of trityl methacrylate units, which may be preferential in one screw sense owing to the sporadic existence of the chiral monomer units. Poly[(S)-α-methylbenzyl methacrylate-co-methacrylic acid] derived from the copolymer by the treatment with methanolic HCl did not show such an abnormal optical rotation.[171]

Table 21. Anionic copolymerization of (S)-α-methylbenzyl methacrylate(M_1) and trityl methacrylate(M_2) by BuLi in THF at $-78\,°C^a$. Ref.[171]

$[M_1]_0/[M_2]_0$ (mol/mol)	Time (min)	Yield (wt%)	m_1/m_2 (mol/mol)	$[\alpha]_{589}^{20\,b}$ (deg·cm³/g·dm)	$[\alpha]$calcd (deg·cm³/g·dm)	Tacticity (%)		
						I	H	S
∞	48 hr	93	∞	−103.8	−103.8	12	28	60
0.25	5	22	0.47	−37.4	−22.2	37	36	27
0.25	95	96	0.23	+38.0	−12.4	57	26	17
0.053	180	96	0.053	+126.4	−3.1	84	11	5

a $[M_1]_0 + [M_2]_0 = 10$ mmol, BuLi $0.2 \sim 0.5$ mmol, THF 20 ml.
b Measured in toluene, C = 0.02 g/cm³.

16 Stereoregularity of Poly(methyl α-phenylacrylate-co-methyl methacrylate)[172, 173]

When methyl α-phenylacrylate and methyl methacrylate were copolymerized by BuLi in toluene, the α-phenylacrylate predominantly polymerized at low temperature, such as −78 °C, to form an atactic homopolymer of this monomer. The α-phenylacrylate has higher reactivity than that of methyl methacrylate because of its lower electron density at the β-carbon and the higher resonance stabilization of its anion. Above the ceiling temperature of the α-phenylacrylate (about 0 °C) essentially the alternating copolymer of these two monomers was obtained from their equimolar mixture. This was attributed to the crossover propagation occurring predominantly over homopolymerization of methyl methacrylate (Table 22). The copolymerization which was investigated by the slow-growth polymerization technique[140] showed that among the homopolymers of each of the comonomers and the alternating co-polymer two or three were formed according to the polymerization temperature: two homopolymers at −78 °C, all three types of polymers at −40 °C, and poly-(methyl methacrylate) and alternating copolymer at 0 °C. The homopolymer of methyl methacrylate was highly isotactic and the alternating copolymer had random tacticity. We can find the multiplicity of active species also in this copolymerization. When THF was used as a solvent the alternating copolymer began to form at lower temperature than in toluene.

Each of the following methacrylates: benzyl, (RS)-α-methylbenzyl, diphenyl-methyl, α-α-dimethylbenzyl, 1,1-diphenylethyl and trityl methacrylate, was copoly-merized with methyl α-phenylacrylate by BuLi in toluene and in THF. The copoly-merizations proceeded almost similarly to the copolymerization of methyl meth-acrylate and the α-phenylacrylate. Alternating copolymers were obtained above 0 °C, but their coisotacticity increased with increasing bulkiness of methacrylate regardless of the copolymerization medium, suggesting that the stereoregulation by the steric interaction between the growing chain end and the incoming monomer overcomes

Table 22. Copolymerization of methyl α-phenylacrylate(M_1) and methyl methacrylate(M_2) by BuLi in toluene and in tetra-hydrofuran[a], Ref.[173]

Temp. (°C)	Alternate sequence in copolymer (mol%)	
	In toluene	In tetrahydrofuran
−78	0	27
−40	31	71
−20	56	82
0	93	100
30	100	100
40	100	100
50	100	−

[a] M_1: 6.8 mmol, M_2: 6.8 mmol, BuLi 0.68 mmol, volume of reaction mixture 10 ml.

the influence of the counter ion at the end in these copolymerizations. The coisotacticity reached to 60 % in the copolymer of trityl methacrylate and methyl α-phenylacrylate. In the copolymerization at low temperature the relative reactivity of methacrylate against the α-phenylacrylate decreased with increasing bulkiness of the former monomer.[174] The optical activity of poly[methyl α-phenylacrylate-co-(S)-α-methylbenzyl methacrylate] was lower than the calculated value and it decreased as the content of alternating sequence increased.[174]

Acknowledgement. The authors are indebted to Dr. Y. Okamoto for his frequent discussions during the preparation of this manuscript. They are also grateful to Mrs. F. Yano for her clerical assistance in preparing manuscript.

17 References

1. Fox, T. G., Garret, B. S., Goode, W. E., Gratch, S., Kincaid, J. F., Spell, A., Stroupe, J. D.: J. Am. Chem. Soc. *80*, 1768 (1958)
2. Miller, R. G. J., Mills, B., Small, P. A., Turner-Jones, A., Wood, D. G. M.: Chem. & Ind.: *1958*, 1323
3. Bovey, F. A., Tiers, G. V. D.: J. Polym. Sci. *44*, 173 (1960)
4. Nishioka, A., Watanabe, H., Yamaguchi, I., Shimizu, H.: J. Polym. Sci. *45*, 232 (1960)
5. Bywater, S.: Adv. Polym. Sci. *4*, 66 (1965)
6. Pino, P., Suter, U. W.: Polymer *17*, 977 (1976)
7. Hatada, K.: Kobunshi *22*, 240 (1973)
8. Hatada, K., Terawaki, Y., Okuda, H., Ohshima, J., Yuki, H.: Kobunshi Kagaku *29*, 391 (1972)
9. Hatada, K., Terawaki, Y., Okuda, H., Niinomi, T., Yuki, H.: Kobunshi Kagaku *28*, 293 (1971)
10. Hatada, K., Ohta, K., Terawaki, Y., Yuki, H.: Kogyo Kagaku Zasshi *71*, 1168 (1968)
11. Hatada, K., Umemura, Y., Furomoto, M., Kokan, S., Ohta, K., Yuki, H.: Makromol. Chem. *178*, 1215 (1977)
12. Hatada, K., Furomoto, M., Kokan, S., Yuki, H.: to be published
13. Cottam, B. J., Wiles, D. M., Bywater, S.: Can. J. Chem. *41*, 1905 (1963)
14. Wiles, D. M., Bywater, S.: Trans. Farad. Soc. *61*, 150 (1965)
15. Glusker, D. L., Stiles, E., Yoncoskie, B.: J. Polym. Sci. *49*, 297 (1961)
16. Goode, W. E., Owens, F. H., Fellman, R. P., Snyder, W. H., Moore, J. E.: J. Polym. Sci. *46*, 317 (1960)
17. Goode, W. E., Owens, F. H., Myers, W. L.: J. Polym. Sci. *47*, 75 (1960)
18. a) Kawabata, N., Tsuruta, T.: Makromol. Chem. *86*, 231 (1965)
    b) Kawabata, N., Tsuruta, T.: Kogyo Kagaku Zasshi *68*, 339 (1965)
19. Wiles, D. M., Bywater, S.: J. Phys. Chem. *68*, 1983 (1964)
20. Yuki, H., Hatada, K., Kokan, S.: Polymer J. *5*, 329 (1973)
21. Hatada, K., Furomoto, M., Umemura, Y., Kitayama, T., Yuki, H.: to be published
22. Amerik, Y., Reynolds, W. F., Guillet, J. E.: J. Polym. Sci., A–1, *9*, 531 (1971)
23. Guzmán, G. M., Bello, A.: Makromol. Chem. *107*, 46 (1967)
24. Rig, A., Figueruelo, J. E., Llano, E.: J. Polym. Sci., B, *3*, 171 (1965)
25. Katchalsky, A., Eisenberg, H.: J. Polym. Sci. *6*, 145 (1951)
26. Hatada, K., Okamoto, Y., Ohta, K., Yuki, H.: J. Polym. Sci. Polym. Lett. Ed. *14*, 51 (1976)
27. Hatada, K., Kitayama, T., Okamoto, Y., Ohta, K., Umemura, Y., Yuki, H.: Makromol. Chem. *178*, 617 (1977)
28. Hatada, K., Kitayama, T., Okamoto, Y., Ohta, K., Umemura, Y., Yuki, H.: Makromol. Chem. *179*, 485 (1978)
29. Hatada, K., Ishikawa, H., Kitayama, T., Yuki, H.: Makromol. Chem. *178*, 2753 (1977)

30. Hatada, K., Ohta, K., Okamoto, Y., Kitayama, T., Umemura, Y., Yuki, H.: J. Polym. Sci., Polym. Lett. Ed. *14*, 531 (1976)
31. Yuki, H., Hatada, K., Kikuchi, Y., Niinomi, T.: J. Polym. Sci. Part B, *6*, 753 (1968)
32. Yuki, H., Hatada, K., Niinomi, T., Kikuchi, Y.: Polymer J. *1*, 36 (1970)
33. Okamoto, Y., Nakashima, S., Ohta, K., Hatada, K., Yuki, H.: J. Polym. Sci., Polym. Lett. Ed. *13*, 273 (1975)
34. Yuki, H., Ohta, K., Hatada, K., Okamoto, Y.: Polymer J. *9*, 511 (1977)
35. Yuki, H., Okamoto, Y., Shimada, Y., Ohta, K., Hatada, K.: J. Polym. Sci., Polym. Chem. Ed. *16*, (1978) in press
36. Harwood, H. J.: Angew. Makromol. Chem. *4/5*, 279 (1968)
37. Matsuzaki, K., Kanai, T., Yamawaki, K., Samre Rung, K. P.: Makromol. Chem. *174*, 215 (1973)
38. Tsuruta, T., Makimoto, T., Kanai, H.: J. Macromol. Sci. *1*, 31 (1966)
39. Nishino, J., Nakahata, H., Sakaguchi, Y.: Polymer J. *2*, 555 (1971)
40. Niezette, J., Desreux, V.: Makromol. Chem. *149*, 177 (1971)
41. Matsuzaki, K., Ishida, A., Takeno, N.: J. Polym. Sci. *C16*, 2111 (1967)
42. Yuki, H., Okamoto, Y., Shimada, Y., Ohta, K., Hatada, K.: Polymer *17*, 618 (1976)
43. Yuki, H., Hatada, K., Ohta, K., Okamoto, Y.: J. Macromol. Sci.,-Chem. *A9*, 983 (1975); Applied Polymer Symposium, No. 26, 39 (1975)
44. Wesslén, B., Gunneby, G., Hellström. G., Svedling, P.: J. Polym. Sci. *C42*, 457 (1973)
45. Casals, P. F., Boccacio, G., Gueniffey, H.: Makromol. Chem. *176*, 2745 (1975)
46. Iwakura, Y., Toda, F., Ito, T., Aoshima, K.: Makromol. Chem. *104*, 26 (1967)
47. Akashi, M., Inaki, Y., Takemoto, K.: Makromol. Chem. *178*, 353 (1977)
48. Fowells, W., Schuerch, C., Bovey, F. A., Hood, F. P.: J. Am. Chem. Soc. *89*, 1396 (1967)
49. Yuki, H., Ohta, K., Ono, K., Murahashi, S.: J. Polym. Sci. A-1, *6*, 829 (1968)
50. Ito, T., Aoshima, K., Toda, F., Uno, K., Iwakura, Y.: Polymer J. *1*, 278 (1970)
51. Aylward, N. N.: J. Polym. Sci. A–1, *8*, 319 (1970)
52. Wiles, D. M., Brownstein, S.: J. Polym. Sci. B, *3*, 951 (1965)
53. Iwakura, Y., Toda, F., Ito, T., Aoshima, K.: J. Polym. Sci. B, *5*, 29 (1967)
54. Yoshino, T., Komiyama, J., Shinomiya, M.: J. Am. Chem. Soc. *86*, 4482 (1964)
55. Yoshino, T., Shinomiya, M., Komiyama, J.: J. Am. Chem. Soc. *87*, 387 (1965)
56. Yoshino, T., Kuno, K.: J. Am. Chem. Soc. *87*, 4404 (1965)
57. Matsuzaki, K., Nishida, Y., Kumahara, H., Miyabayashi, T., Yasukawa, T.: Makromol. Chem. *167*, 139 (1973)
58. Maruhashi, M., Takida, H.: Makromol. Chem. *124*, 172 (1969)
59. Trekoval, J., Lim, D.: J. Polym. Sci. *C4*, 333 (1964)
60. Trekoval, J.: J. Polym. Sci. A–1, *9*, 2575 (1971)
61. Trekoval, J., Vlček, P.: Preprints of IUPAC Congress on Macromolecules July, 1977, Dublin Ireland p.55
62. Vlček, P., Trekoval, J.: Makromol. Chem. *176*, 2595 (1975)
63. Klippert, H., Ringsdorf, H.: Makromol. Chem. *153*, 289 (1972)
64. Braun, D., Herner, M., Johnsen, U., Kern, W.: Makromol. Chem. *51*, 15 (1962)
65. Roig, A., Figueruelo, J. E., Llano, E.: J. Polym. Sci. *C16*, 4141 (1968)
66. Junquera, J., Cardona, N., Figueruelo, J. E.: Makromol. Chem. *160*, 159 (1972)
67. Pascault, J. P., Kawak, J., Golé, J., Pham, Q. T.: Europ. Polym. J. *10*, 1107 (1974)
68. Allen, K. A., Gowenlock, B. G., Lindsell, W. E.: J. Polym. Sci., Polym. Chem. Ed. *12*, 1131 (1974)
69. Alev, S., Schué, F., Kaempf, B.: J. Polym. Sci., Polym. Lett. Ed. *13*, 397 (1975)
70. Kaempf, B., Raynal, S., Lacoste, J., Schué, F.: J. Polym. Sci., Polym. Lett. Ed. *12*, 211 (1974)
71. Viguier, M., Abadie, M., Schué, F., Kaempf, B.: Europ. Polym. J. *13*, 213 (1977)
72. Pascault, J. P., Chastrette, F., Pham, Q. T.: Europ. Polym. J. *12*, 273 (1976)
73. Müller, A. H. E., Höcker, H., Schulz, G. V.: Macromolecules *10*, 1086 (1977)
74. Volker, T., Neumann, A., Baumann, M.: Makromol. Chem. *63*, 182 (1963)
75. Freireich, S., Zilkha, A.: J. Macromol. Sci.,-Chem. *A6*, 1383 (1972)

 76. Shimomura, T., Ono, K., Tsuchida, E., Shinohara, I.: Kogyo Kagaku Zasshi *71*, 1070 (1968)
 77. Tomoi, M., Sekiya, K., Kakiuchi, H.: Polymer J. *6*, 438 (1974)
 78. Fujishige, S., Suzuki, M., Cantow, H. –J.: Preprints of the 2nd Japan-USSR Polymer Symposium, Kyoto, 1976, p.99
 79. Nishioka, A., Watanabe, H., Abe, K., Sono, I.: J. Polym. Sci. *48*, 241 (1960)
 80. Watanabe, H.: Kogyo Kagaku Zasshi *65*, 1104 (1968)
 81. Yamada, A., Yanagita, M., Hirose, M.: Kogyo Kagaku Zasshi *74*, 1185 (1971)
 82. Ando, I., Chujo, R., Nishioka, A.: Polymer J. *1*, 609 (1970)
 83. Okamoto, Y., Urakawa, K., Yuki, H.: Polymer J., *10*, 457 (1978)
 84. Breslow, D. S., Kutner, A.: J. Polym. Sci., B, *9*, 129 (1971)
 85. Joh, Y., Kotake, Y.: Macromolecules *3*, 337 (1970)
 86. Dever, G. R., Karasz, F. E., MacKnight, W. J., Lenz, R. W.: J. Polym. Sci., Polym. Chem. Ed. *13*, 2151 (1975)
 87. Uryu, T., Ohaku, K., Matsuzaki, K.: J. Polym. Sci., Polym. Chem. Ed. *12*, 1723 (1974)
 88. Allen, P. E. M., Bateup, B. O.: J. Chem. Soc. Faraday Transactions I, *71*, 2203 (1975)
 89. Chlandra, F. P., Donaruma, L. G.: J. Appl. Polym. Sci. *15*, 1195 (1971)
 90. Murahashi, S., Niki, T. Obokata, T., Yuki, H., Hatada, K.: Kobunshi Kagaku *24*, 198 (1967)
 91. Murahashi, S., Obokata, T., Yuki, H., Hatada, K.: Kobunshi Kagaku, *24*, 309 (1967)
 92. Abe, H., Imai, K., Matsumoto, M.: J. Polym. Sci. *C23*, 469 (1968)
 93. Abe, H., Imai, K., Matsumoto, M.: J. Polym. Sci. B, *3*, 1053 (1965)
 94. Shirokov, N. A., Mazurek, V. V.: Polymer J. *7*, 523 (1975)
 95. Yuki, H., Hatada, K., Niinomi, T., Hashimoto, M., Ohshima, J.: Polymer J. *2*, 629 (1971)
 96. Natta, G., Mazzanti, G., Longi, P., Bernardini, F.: Chim. Ind.(Milano) *42*, 457 (1960)
 97. Ikeda, M., Hirano, T., Tsuruta, T.: Makromol. Chem. *150*, 127 (1971)
 98. Tomoi, M., Kurita, H., Onozawa, M., Kakiuchi, H.: Nippon Kagaku Kaishi *1976*, 356
 99. Abe, H., Imai, K., Matsumoto, M.: J. Polym. Sci., B, *4*, 589 (1966)
100. Abe, H., Imai, K., Matsumoto, M.: J. Polym. Sci., B, *4*, 861 (1966)
101. Dixit, S. S., Deshpande, A. B., Kapur, S. L.: Europ. Polym. J. *7*, 699 (1971)
102. Chikanishi, K., Tsuruta, T.: Makromol. Chem. *81*, 198 (1965)
103. Tsuruta, T., Kawakami, Y., Tsushima, R.: Makromol. Chem. *149*, 135 (1971)
104. Yuki, H., Hatada, K., Niinomi, T., Miyaji, K.: Polymer J. *1*, 130 (1970)
105. Hatada, K., Kokan, S., Niinomi, T., Miyaji, K., Yuki, H.: J. Polym. Sci., Polym. Chem. Ed. *13*, 2117 (1975)
106. Wesslen, B., Lenz, R. W.: Macromolecules *4*, 20 (1971)
107. Wesslen, B., Lenz, R. W., MacKnight, W. J., Karasz, F. E.: Macromolecules *4*, 24 (1971)
108. Wesslen, B., Lenz, R. W., Bovey, F. A.: Macromolecules *4*, 709 (1971)
109. Dever, G. R., Karasz, F. E., MacKnight, W. J., Lenz, R. W.: J. Polym. Sci., Polym. Chem. Ed. *13*, 1803 (1975)
110. Uryu, T., Matsuzaki, K.: J. Polym. Sci., B, *10*, 867 (1972)
111. Gisser, H., Mertwoy, H. E.: Macromolecules *7*, 431 (1974)
112. Stroupe, J. D., Hughes, R. D.: J. Am. Chem. Soc. *80*, 2341 (1958)
113. Liquori, A. M., Anzuino, G., D'Alagni, M., Vitagliano, V., Costantino, L.: J. Polym. Sci. A–2, *6*, 509 (1968)
114. Buter, R., Tan, Y. Y., Challa, G.: Polymer *14*, 171 (1973)
115. Inagaki, H., Kamiyama, F.: Macromolecules *6*, 107 (1973)
116. Miyamoto, T., Tomoshige, S., Inagaki, H.: Polymer J. *6*, 564 (1974)
117. Hatada, K., Umemura, Y., Yuki, H.: Preprints of IUPAC Congress on Marcromolecules, July, 1977, Dublin Ireland p.63
118. Hatada, K., Kitayama, T., Umemura, Y., Furomoto, M., Yuki, H.: to be published
119. Warzelhan, V., Schulz, G. V.: Makromol. Chem. *177*, 2185 (1976)
120. Koide, N., Iimura, K., Takeda, M.: J. Macromol. Sci.,-Chem. *A9*, 961 (1975)
121. Bateup, B. O., Allen, P. E. M.: Europ. Polym. J. *13*, 761 (1977)
122. Buter, R., Tan, Y. Y., Challa, G.: J. Polym. Sci., A–1, *10*, 1031 (1972)
123. Buter, R., Tan, Y. Y., Challa, G.: J. Polym. Sci., Polym. Chem. Ed. *11*, 1003 (1973)
124. Buter, R., Tan, Y. Y., Challa, G.: J. Polym. Sci., Polym. Chem. Ed. *11*, 1013 (1973)

125. Buter, R., Tan, Y. Y., Challa, G.: J. Polym. Sci., Polym. Chem. Ed. *11*, 2975 (1973)
126. Gons, J., Vorenkamp, E. J., Challa, G.: J. Polym. Sci., Polym. Chem. Ed. *13*, 1699 (1975)
127. Gons, J., Slagter, W. O., Challa, G.: J. Polym. Sci., Polym. Chem. Ed. *15*, 771 (1977)
128. Miyamoto, T., Tomoshige, S., Inagaki, H.: Makromol. Chem. *176*, 3035 (1975)
129. Spěváček, J., Schneider, B.: Makromol. Chem. *175*, 2939 (1974)
130. Spěváček, J., Schneider, B.: Makromol. Chem. *176*, 729 (1975)
131. Spěváček, J., Schneider, B.: Makromol. Chem. *176*, 3409 (1975)
132. Hatada, K., Terawaki, Y., Furomoto, M., Kitayama, T., Yuki, H.: unpublished data
133. Coleman, B. D., Fox, T. G.: J. Chem. Phys. *38*, 1065 (1963)
134. Coleman, B. D., Fox, T. G.: J. Am. Chem. Soc. *85*, 1241 (1963)
135. Hatada, K., Okamoto, Y., Ise, H., Yamaguchi, S., Yuki, H.: J. Polym. Sci., Polym. Lett. Ed. *13*, 731 (1975)
136. Nozakura, S., Okamoto, T., Toyora, K., Murahashi, S.: J. Polym. Sci., Polym. Chem. Ed. *11*, 1043 (1973)
137. Nozakura, S., Ishihara, S., Inaba, Y., Matsumura, K., Murahashi, S.: J. Polym. Sci., Polym. Chem. Ed. *11*, 1053 (1973)
138. Hatada, K., Kameyama, Y., Sugino, H., Yuki, H.: unpublished data
139. Asami, R.: Preprints of the 22nd Annual Meeting of the Society of Polymer Science, Japan, 1973, 31C25
140. Hatada, K., Kokan, S., Yuki, H.: J. Polym. Sci., Polym. Lett. Ed. *13*, 721 (1975)
141. Hatada, K., Furomoto, M., Yuki, H.: Makromol. Chem. *179*, 1107 (1978)
142. Schildknecht, C. E., Zoss, A. O., Grosser, F.: Ind. Eng. Chem. *41*, 2891 (1949)
143. Hatada, K., Furomoto, M., Yuki, H.: to be published
144. Matsuzaki, K., Tateno, N.: J. Polym. Sci. *C23*, 733 (1968)
145. Solomatina, I. P., Aliev, A. D., Krentsel, B. A.: Vysokomol. soed. *A11*, 871 (1969)
146. Ikeda, M., Hirano, T., Nakayama, S., Tsuruta, T.: Makromol. Chem. *175*, 2775 (1974)
147. Okamoto, Y., Urakawa, K., Ohta, K., Yuki, H.: Macromolecules, *11*, 719 (1978)
148. Okamoto, Y., Ohta, K., Yuki, H.: Chem. Lett. *1977*, 617 (1977)
149. Okamoto, Y., Ohta, K., Yuki, H.: Macromolecules, *11*, 724 (1978)
150. Krause, S., Gormley, J. J., Roman, N., Shetter, J. A., Watanabe, W. H.: J. Polym. Sci., A, *3*, 3573 (1965)
151. Gargallo, L., Russon, M.: Makromol. Chem. *176*, 2735 (1975)
152. Karasz, F. E., MacKnight, W. J.: Macromolecules *1*, 537 (1968)
153. Dever, G. R., Karasz, F. E., MacKnight, W. J., Lenz, R. W.: Macromolecules *8*, 439 (1975)
154. Thompson, E. V.: J. Polym. Sci., A–2, *4*, 199 (1966)
155. Heatley, F., Begum, A.: Polymer *17*, 399 (1976)
156. Lyerla, J. R. Jr., Horikawa, T. T.: J. Polym. Sci., Polym. Lett. Ed. *14*, 641 (1976)
157. Lyerla, J. R. Jr., Horikawa, T. T., Johnson, D. E.: J. Am. Chem. Soc. *99*, 2463 (1977)
158. Spěváček, J., Schneider, B.: Polymer *19*, 63 (1978)
159. Otsu, T., Ito, T., Imoto, M.: J. Polym. Sci. *C16*, 2121 (1967)
160. Yuki, H., Okamoto, Y., Ohta, K., Hatada, K.: Polymer J. *6*, 573 (1974)
161. Yuki, H., Okamoto, Y., Ohta, K., Hatada, K.: J. Polym. Sci. Polym. Chem. Ed. *13*, 1161 (1975)
162. Vlček, P., Doskočilova, D., Trekoval, J.: J. Polym. Sci. *C42*, 231 (1973)
163. Ito, K., Sugie, T., Yamashita, Y.: Makromol. Chem. *125*, 291 (1969)
164. Bevington, J. C., Harris, D. O., Rankin, F. S.: Europ. Polym. J. *6*, 725 (1970)
165. Ohta, K., Yuki, H., Okamoto, Y., Hatada, K.: J. Polym. Sci. Polym. Chem. Ed. in press
166. Yuki, H., Hatada, K., Ohta, K., Okamoto, Y.: Preprints of Symposium on the Mechanism and Active Species in Ionic Polymerization, Nagoya, October, (1975) p.199
167. Klesper, E., Gronski, W.: J. Polym. Sci. B, *7*, 661 (1969)
168. Klesper, E., Gronski, W.: J. Polym. Sci. B, *7*, 727 (1969)
169. Yamada, A., Tanaka, J.: Preprints of 23rd Symposium on Macromolecular Science, Japan, Tokyo, 1974
170. Yuki, H., Ohta, K., Okamoto, Y., Hatada, K.: Polymer J. *10*, 505 (1978)

171. Yuki, H., Ohta, K., Okamoto, Y., Hatada, K.: J. Polym. Sci., Polym. Lett. Ed. *15*, 589 (1977)
172. Yuki, H., Hatada, K., Ohshima, J., Komatsu, T.: Polymer J. *2*, 812 (1971)
173. Hatada, K., Ohshima, J., Komatsu, T., Kokan, S., Yuki, H.: Polymer *14*, 565 (1973)
174. Yuki, H., Ohta, K., Hatada, K., Ishikawa, H.: Polymer J. to be submitted

Received July 27, 1978
T. Saegusa (editor)

Molecular Sieves as Polymerization Catalysts

Mukul Biswas and Narayan C. Maity

Department of Chemistry, Indian Institute of Technology, Kharagpur 721302, India

Molecular sieves have potential uses as catalysts for vinyl polymerization reactions. They have been widely used as cationic catalysts, and in a limited number of instances in anionic polymerization reactions. Molecular sieves also find uses as polymer modifiers bringing in definite improvements in their properties.

Table of Contents

I Introduction

Molecular sieves are synthetic crystalline aluminosilicates, honeycombed with cavities which are interconnected by pores varying from about 3 to 10 Å units in diameter. The basic structural units of these sieves are Si and Al atoms, tetrahedrally coordinated with four oxygen atoms. The oxygen atoms are mutually shared between tetrahedron units contributing one of the two valence charges of each oxygen atom to each tetrahedron. Since Al atoms are trivalent each AlO_4^- is negatively charged and the charge on these units is balanced by cations, usually Na^+ and K^+ ions. These cations are exchangeable with other cations accounting for base exchange properties of the molecular sieves.

An important building block of these molecular sieves is the sodalite cage, a truncated octahedron unit, consisting of 24 (Si,AlO_4) units. In type A zeolites the sodalite cages are joined together by four membered rings. In types X and Y zeolites the sodalite cages are joined together by six membered rings.

The pores of a particular type of sieve are precisely uniforms in size[1], as revealed in Fig. 1. Adsorption selectivity based on molecular size together with a selective preference for polar or polarizable molecules make these high internal surface area materials outstanding adsorbents.

Molecular sieve catalysts are characterized by high activity and good selectivity in a wide variety of reactions. Their versatility in a variety of organic reactions, such as skeletal isomerization, selective hydrogenation, dehydration and dehydrogenation, as well as the esterification of acids and the dehydrohalogenation of halogen containing compounds has been well established. Another potential field of application of these molecular sieves is as polymerization catalysts and by now there is a considerable accumulation of literature on this aspect. The purpose of this review is to highlight the significant developments in this field, and to present a coherent and up-to-date account of the general kinetic mechanistic and related aspects of the different polymerization reactions catalysed by the molecular sieves. Tables 1 and 2 list the molecular sieves used in polymerization and post polymerization reactions.

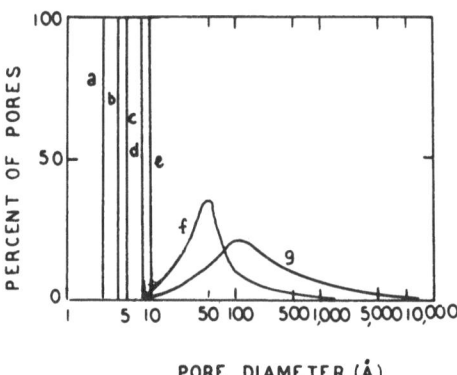

Fig. 1. Pore size distribution of various molecular sieves. (a) Linde molecular sieve type 3 A; (b) type 4 A; (c) type 5 A; (d) type 10 X; (e) type 13 X; (f) silica gel; (g) activated carbon[1]

PORE DIAMETER (Å)

Table 1. Tabular survey on the molecular sieves used as polymerization catalyst

Molecular sieves	Monomer	Ref.
Acidic alumina silicatezeolite catalyst	Propylene	32)
Eroinite	Propylene	31)
Chabasite	Butylene	34)
	t-Butyl chloride and	
	Butyl bromide	35, 36)
Ca^{2+} form Y-zeolite	Ethylene	20)
Ca-A zeolite	Tetrafluoroethylene	24)
Ca-Z type zeolite (Co^{2+} substituted)	Propylene	28)
CaNa Y-zeolite (1% Pd exchanged)	Ethylene	21)
Calsit 5+ KOH+N-Benzoyl caprolactam	ω-Caprylolactam	57)
Cationised zeolite (containing elements of group I, II, III, and IV	Ethylene	19)
CoX	Propylene	29)
Co modified and nonmodified NaY catalyst	1-Propyl naphthalene	55)
Cr^{3+} type of zeolite	Ethylene	20)
Cr-Y zeolite	Ethylene	23)
H-mordenite	n-Butyl vinyl ether	11)
Hydrogen faujasite (H−Y)	Ethyl and Isobutyl vinyl ether	12)
Hydrogen form of mordenite	Propylene	31)
H-form alumino silicate	Butadiene	54)
H−Y	Ethylene, Propylene	15)
	Hex-l-ene, 2,3, Dimethyl but -1-ene	18)
Na zeolite	Alkylvinyl ether	10)
NaY	Tetrafluoroethylene	24)
Nax (Co^{2+} substituted)	Propene	28)
NaHY	Propene	33)
Nalsit 4A + NaOH+N-acetylcaprolactam	ε-Caprolactam	56)
Na-alumino silicate	2-Alkoxy propene	49)
Nax and NaY (exchanged with Ni^{2+}, Cu^{2+}, Nd^{3+})	Isobutylene	47, 48)
NdNax	Propylene	30)
Ni-X	Propylene	29)
Organo clay catalyst having ⩾ 1 aminogroup	Caprolactam	58)
Pd-zeolite	Isoprene and acrylonitrile	52)
Rare earth X zeolites	Ethylene, propylene, hex-l-ene, 2,3-dimethyl but-l-ene	18)

Table 1. (continued)

Molecular sieves	Monomer	Ref.
SK 40 modified with Cr^{3+}	Ethylene	22)
SK 45, SK 100, SK 110, SK 200, SK 310, SK 400, SK 410, SK 500	Isobutylene	40)
XY mordenite (in the H-form)	Styrene	9)
Zeolite 3A	Styrene	2)
	Olefin[a]	25, 26)
	Isobutylene	40)
Zeolite 4A	Styrene	2)
	Olefin[a]	25, 26)
	Isobutylene	40)
	2-Alkoxy propene	49)
	2,2-Dimethyl-4-methylene-1,3-dioxolane	50)
	2-Methyl-4-methylene-1,3-dioxolane	50)
	2,6-Dimethyl phenol	51)
Zeolite 5A	Styrene	2)
	Ethyl vinyl ether	2)
	Olefin[a]	25, 26)
	Isobutylene	37, 39)
	Isobutylene	40)
	2,2-Dimethyl-4-methylene-1,3-dioxolane	50)
	2-Methyl-4-methylene-1,3-dioxolane	50)
Zeolite 10X	Styrene	2)
	Ethyl vinyl ether	2)
	Olefin[a]	25, 26)
	Isobutylene	40)
Zeolite 13X	Styrene	2)
	Ethyl vinyl ether	2)
	N-vinyl carbazole	13)
	Olefin[a]	25, 26)
	Isobutylene	40)
Zeolite 13X containing Ni^{2+}	Propylene	27)

[a] Ethylene, propylene, isobutylene.

Table 2. Miscellaneous applications of molecular sieves in polymerization and post polymerization reaction

Molecular sieves	Application	Ref.
CaX	Vulcanization agent for chloroprene rubber	59)
	Improvement of the compactness of the rubber product	64)
CaA	Vulcanization agent for chloroprene rubber	59)
MgA	Vulcanization agent for chloroprene rubber	59)
Molecular sieves treated with organic compounds containing oxygen in the form of OH group or ether linkages	Inhibition of polymerization	67)
NaA	Vulcanization agent for chloroprene rubber	60)
NaX	Crosslinking of high density poly ethylene	61)
NH_4A	Vulcanization agent for chloroprene rubber	59)
Zeolite modified with organic solvents	Increase of the resistance and degree of radiation crosslinking of filled poly vinyl chloride	62)
Zinc salt of molecular sieve	Increase of the molecular weight of polyester and polyamide	63)
Zeolite	Improvement of antistatic properties of polymer	66)
Zeolite 13X, 5A, 4A, 10X XW and also natural zeolites	Improvement of antistatic properties of polymer	65)

II Cationic Polymerization of Vinyl Monomers

A. Styrene and Its Derivatives

The first significant work involving the cationic polymerization of styrene and its derivatives (α-methyl styrene, p-chlorostyrene) on molecular sieves appears to have been due to Panaiotov and Dimitrov[2] in 1967. Molecular sieves, 3A,4A,5A,10X and 13X (200 mesh) after activation at 400 °C for 15 minutes in a current of dry and pure nitrogen polymerize vinyl monomers in bulk or in dichloroethane solvent. The yield data for polystyrene polymer with varying molecular sieves (Table 3), different preheating temperatures for a particular catalyst 10X (Table 4) and the

Table 3. Dependence of the yield of polystyrene on the type of molecular sieves[a]

Molecular sieve	Yield of polystyrene[b] (%)	
	Without solvent	With solvent
3A	0.4	0.9
4A	0.7	5.5
5A	36.0	55.0
10X	55.0	60.0
13X	20.0	2.0

[a] Cited from Ref.[2].
[b] Polymerization done at room temperature.

Table 4. Dependence of catalytic activity[a] of the molecular sieves[b] on its thermal processing

Temperature of heating the catalyst before polymerization ($^\circ$C)	Yield of polymer (%)
Not heated	1
100	12
200	60
300	55–60
400	50

[a] Reproduced from Ref.[2].
[b] 10X molecular sieves preheated for 15 minutes at the temperature shown. Bulk polymerization carried out at room temperature.

Table 5. Molecular weight[a] of polystyrene, obtained under different conditions[b]

Molecular sieve	Temperature of polymerization ($^\circ$C)	Solvent	Molecular weight
4A	Room	Dichloroethane	6.000
5A	Room	Dichloroethane	8.000
5A	−60	Dichloroethane	12.000
5A	Room	Without solvent	24.700
10X	Room	Dichloroethane	8.000
10X	−60	Dichloroethane	15.400
10X	Room	Without solvent	24.000

[a] Determined viscometrically in benzene at 20 $^\circ$C [η] = 1.23 x 10^{-4} $M^{0.72}$.
[b] Reproduced from Ref. [2].

molecular weight data under different conditions (Table 5) allow the following conclusions to be drawn on the system.

(i) The polymerization activity of the various sieves decreases in the order, $3A < 4A < 5A < 10X < 13X$.

(ii) A preheating temperature corresponding to $200\,^{\circ}C$ furnishes the optium polymerization yield of 60% with 10X sieve in absence of solvent at room temperature.

(iii) Molecular weights of the polymer obtained by bulk polymerization are greater than those obtained by solution polymerization. Temperature favours higher molecular weight formation.

Interestingly, the polystyrene produced by solution polymerization is neither isotactic nor atactic, no crystallinity being detected in the product polymer under these conditions.

A plausible model for explaining the polymerization involves the formation of an active centre in presence of water as a cocatalyst.

Such an active centre is capable of causing cationic polymerization whereby an active end of the growing chain is fixed on the hard surface of the catalyst. Schematically,

growth of the chain,

Termination reasonably occurs by the conventional path.

The trend in the activity of the molecular sieves reveals that 5A and 10X containing Ca^{+2} ions are more active than those containing Na^+ ions. This difference is attributed to the distribution of the electronic densities of the corresponding complex counter ions. The presence of two positive charges on the calcium cations leads to the weakening of the interaction of the carbon ion in the growing chain with hydroxyl ions to a more appreciable extent than in the case of sodium ions. As a consequence, the calcium containing sieves turn out to be a more active catalyst than the sodium form.

The activity of the cations in one of the main factors which determine the rate of growth of the chain. The molecular weight of the polymers formed under these conditions should be proportional to the oxidizing centre, which forms a catalytic complex with the cocatalyst. Accordingly the molecular weight of the polystyrene over the 4A (sodium form sieve) is expected to be lower than the same with 5A and 10X sieves (calcium form) which is actually endorsed by the Table 5.

The complex counter ion formed on the surface of the molecular sieve possesses a scattered charge and large volume. This may be the reason for the absence of any stereospecificity in the product polymer despite the presence of heterogeneous polarized surface.

The cationic polymerization of styrene was also studied in details by Barson et al.[3] by 13X zeolite at 30 °C. Rate of monomer consumption were followed dilatometrically and the molecular weight distribution of the product was investigated by gel-permeation chromatography.

The observed decrease in the dilatometric reading is seen to be linear over most of the time range. Initially the rate of monomer consumption is high due to the adsorption of styrene on the zeolite. A decrease in the rate of contraction to zero rate observed after *ca.* 30 hours is probably caused by the build up of polymer around the active sites, impeding the diffusion of monomer to these sites and also causing the zeolite particles to adhere together.

The effect of the weight of zeolite on the rate R_M of consumption of monomer, using 13X zeolite, preheated to 290–300 °C in undiluted styrene (Fig. 2a) is described by the equation,

$$R_M \, \alpha \, (\text{weight of zeolite})^{0.55 \pm 0.11}$$

R_M values are substantially lower in the absence of stirring than when stirring is carried out, indicating that diffusion is significant under these conditions.

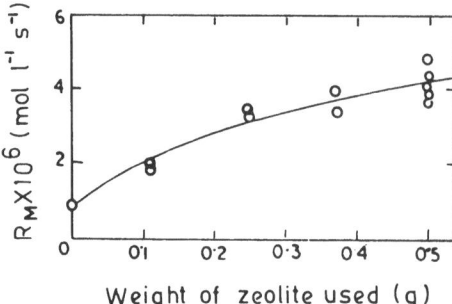

Weight of zeolite used (g)

Fig. 2a. Dependence of rate of styrene consumption on weight of 13 X zeolite[3]

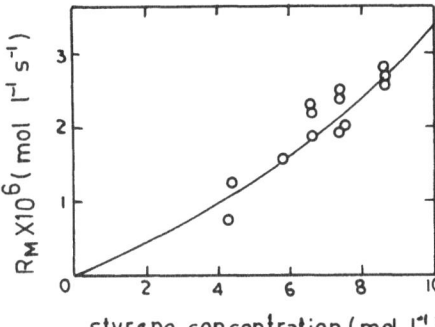

styrene concentration (mol l⁻¹)   **Fig. 2b.** Dependence of rate styrene consumption on the styrene concentration[3]

However, assuming R_M to be the sum of the rate of thermal polymerization $(R_{m,t})$ and of the rate of polymerization on the zeolite surface R_{MZ}, the latter quantity obeys the expression

$$R_{MZ} \; \alpha \; (\text{weight of zeolite})^{0.81 \, \pm \, 0.14}$$

Rychly and Lazar[4] demonstrated that oligostyryl radicals generated from styrene by benzoyl peroxide are stabilised sufficiently by 13X zeolite to be detected by ESR at temperatures upto 50 °C. This would indicate that 13X zeolite could inhibit or retard the radical polymerization of styrene at 30 °C. The thermal polymerization of styrene is believed to be a radical reaction[5]. Hence it may be assumed that the presence of zeolite exerts no effect upon the rate of thermal polymerization.

The effect (Fig. 2b) of the concentration of styrene [M] in cyclohexane on R_M is described by

$$R_M \; \alpha \; [M]^{1.47 \, \pm \, 0.19}$$

and assuming thermal polymerization to be unaffected by the presence of zeolite,

$$R_{MZ} \; \alpha \; [M]^{1.56 \, \pm \, 0.31}$$

As regards the dependence of R_M on the preheating temperature of the zeolite, it has been observed in general the higher the preheating temperature the lower is the value of R_M. It may be noted that 0.1145 ± 0.0032 gm of liquid per gm of zeolite is desorbed on heating the zeolite to 290 ° − 300 °C for 6 hours and on heating to 340 °− 350 °C for 6 hours 0.1406 ± 0.0052 gm of liquid per gm of zeolite is desorbed.

Molecular Weight Trends. The dependences of $\bar{M}_n$ and dispersion on several experimental parameters are presented in Table 6. The data allow the following conclusions.

There is no significant change in $\bar{M}_n$, with increase in the weight of zeolite. The effect of monomer concentration on $\bar{M}_n$ using zeolite preheated to 340 ° to 350 °C is summarized in the following empirical equation,

$$(\bar{M}_n)^{-1} = (1.75 \pm 0.30) \times 10^{-6} \left(\frac{[M]}{\text{mol., lit.}^{-1}} \right)^{-1} + (1.05 \pm 0.19) \times 10^{-7}$$

The molecular weight also depends on the preheating temperature of zeolite. The use of 13X zeolite preheated to 290–300 °C results in the formation of polymers of $\bar{M}_n$ $(1.9 \pm 0.14) \times 10^6$ where as polymers with average $\bar{M}_n$ of $(3.6 \pm 0.3) \times 10^6$ have been obtained with sieves preheated to $ca.$, 340–350 °C.

In almost all cases, the values of the dispersion are less than 1.5 and these values are independent of the weights of zeolite, concentration of the monomer, and the preheating temperature.

Table 6. Variation of number average molecular weight $\bar{M}_n$ and dispersion D with conditions of polymerization[a]

Weight of zeolite (g)	Concentration of styrene (mol 1^{-1})	Duration of experiment (h)	Preheating temperature of zeolite (°C)	$\bar{M}_n \times 10^{-6}$	Dispersion D
0	8.71	30	– – –	1.8	1.5
0.50	8.71	8	290–300	1.3	1.6
0.50	8.71	8	290–300	1.4	1.5
0.50	8.71	8	290–300	1.5	1.5
0.50	8.71	50	290–300	1.7	1.4
0.50	8.71	8	290–300	1.0	1.8
0.50	8.71	8	290–300	2.6	1.3
0.50	8.71	8	290–300	2.3	1.4
0.50[b]	8.71	30	290–300	1.4	1.6
0.50[b]	8.71	30	290–300	1.8	1.5
0.50	8.71	30	290–300	3.0	1.3
0.375	8.71	8	290–300	3.6	1.2
0.375	8.71	8	290–300	2.6	1.5
0.25	8.71	8	290–300	2.1	1.4
0.25	8.71	8	290–300	2.1	1.4
0.125	8.71	8	290–300	2.8	1.4
0.125	8.71	8	290–300	3.2	1.3
0.75[b]	8.71	8	290–300	2.6	1.4
0.50	4.41	30	340–350	1.0	2.6
0.50	4.40	30	340–350	2.0	1.1
0.50	6.72	30	340–350	3.1	1.3
0.50	6.72	30	340–350	2.5	1.4
0.50	7.45	30	340–350	3.4	1.3
0.50	7.51	30	340–350	3.1	1.3
0.50	7.51	30	340–350	2.5	1.3
0.50	7.63	30	340–350	2.6	1.3
0.50	8.71	4	340–350	3.8	1.2
0.50	8.71	30	340–350	3.3	1.3
0.50	8.71	30	340–350	3.8	1.3
0.50	8.71	30	340–350	3.6	1.3

a   Reproduced from Ref.[3], by permission of Blackwell Scientific Publications Ltd.
b   Dilatometer mixture not stirred.

Mechanism. The increase in the rate of monomer consumption with increase in the weight of zeolite suggests that the zeolite must be involved in the initiation or propagation stages. The rate of monomer consumption decreases with increase in the preheating temperatures. As the desorbed liquid is mainly water, the observation can be explained if water plays an important role in the initiation step. Bertsch and Habgood[6] have suggested, on the basis of i. r. studies of NaX type of sieves, that isolated water molecules are absorbed simultaneously by means of ion-dipole interaction with the associated cation and by hydrogen bonding by one of the hydrogen atoms to an oxygen atom on the zeolite surface.

Hirschler[7] suggests that this interaction would give a hydrogen atom of the water molecule an acidic character. The zeolite would thus be considered a proton donor in presence of water. Initiation of the polymerization by proton donation would be consistent with the observation. The decrease in the consumption rate of monomer with increased preheating temperature can also be explained on this basis. The sites activated by water decrease as the temperature increases.

The polymer chains initiated at the zeolite surface by a cationic mechanism would be held in contact with the surface by intermolecular interactions. Under these conditions, the propagation step would also probably be on the zeolite surface. Termination would be likely to occur by dissociation from the zeolite surface producing an unsaturated chain end. The activated zeolite site would also be reformed. This mechanism is somewhat similar to that used by Benson and co-workers[8] for polymerization by acid clay catalysts.

Examination of the results for undiluted monomer (Table 6) indicates that there is no significant change in $\bar{M}_n$ with conversion. This feature is consistent with the existence of termination/transfer reaction. On the contrary, the dispersions of the polymers are < 1.5 and low dispersion values are believed to be associated with reactions having rapid initiation and no termination or transfer.

The zeolite surface is considered to be uniformly scattered with activated sites at each of which a polymer chain can be initiated. At short chain lengths each growing polymer chain can be considered to be isolated from any other chain. There may be a slow decrease (though not marked) in the rate of propagation due to previously formed parts of the polymer chain obstructing the growth point at the activated site. Assuming that the rate of termination is low compared to the rate of propagation, the probability of termination at any given chain length will also be low. Finally, the polymer molecules grow to a size such that they occupy all the available surface area of the zeolite. At this stage, the rate of propagation at any one site decreases more rapidly due to the increase in thickness and density of the polymer deposits around each activated site. Increased rate of termination would be likely at this stage, since there would be an increased strain between the growing chain end and the chain in the vicinity of the activated site. Removal of the terminated chains from the zeolite surface would allow for possible initiation of further chains, which, however, would proceed with more difficulty as the reaction progresses more and more.

On the basis of these observations the molecular size of the polymers would be controlled by the surface area of the zeolite per activated site, and also there would be a narrower molecular weight distribution. The molecular weights of the polymers would also not be expected to depend on the weight of the zeolite. With increase in the monomer concentration the rate of propagation would increase relative to the rate of termination.

Among other sieves XY mordenite in the H form initiate[9] the solution poly-merization of styrene. Interestingly, the system is believed to result in the for-mation of polymeric molecular sieves with altered ion exchange and molecular adsorption characteristics.

B. Vinyl Ethers

Ethyl vinyl ether[2] undergoes exothermic polymerization in presence of 3A, 4A, 5A, 10X and 13X molecular sieves. Polyvinylethers ranging from viscous oil to hard solid have been insolated through the polymerization (-70 to $150\,°C$) of alkyl-vinylethers over sodium zeolite, the lower temperatures favouring higher molecular weights[10]. With a typical recipe consisting of 15 g of 5A molecular sieve (Linde) preheated to $315\,°C$ for 3 hours and 200 ml of isobutylvinylether, 3% yield of colourless sticky polymer $[\eta]_{toluene} = 0.29$ has been obtained in course of 18 hours[10]. However, some what comprehensive studies are due to Barrer and Oei[11, 12] on the polymerization of the ethyl, n-nutyl and iso-butyl vinyl ether over H-mordenite and H-faujasite (H-Y). The H-mordenite is available as H-zeolon. The H-Y may be prepared from NaY by the reaction

$$\text{Na−Y} \xrightarrow[\text{1 day}]{\text{4.24 M NH}_4\text{Cl (250 ml)}} \text{NH}_4\text{−Y} \xrightarrow[\text{3 days in air}]{360\,°C} \text{H−Y}$$
$$(100\,\text{g})$$

The cylinder shaped H-mordenite crystals are characterized by average dia-meters and lengths of 4.2 and $4.6\,\mu m$ respectively, and the cubic crystals of H-Y possess average edge lengths of $1.35\,\mu m$.

The polymerization was followed with silica spring balances, each balance and manometer being housed in separate air thermostat ($\pm 0.2\,°C$). The zeolite samples were out-gassed for 7 days at temperatures rising to $360\,°C$ and placed in small glass buckets suspended from the calibrated silica springs. The n-butyl vinyl ether monomer was freshly distilled before admission as vapour from a reservoir through the vacuum line to the balance cases. The amount of water in the zeolite was altered in steps from zero upwards by admitting the regulated amount of water to the zeolite at $30\,°C$ followed by heating the zeolite to $ca.,135\,°C$ for 24 hours and subsequent slow cooling the sample at $30\,°C$ − a procedure which helped the intra-crystalline water to be distributed as uniformly as possible.

Kinetics. Hydrogen mordenite appears to be a good catalyst for the poly-merization of n-butylvinylether. At $30\,°C\ Q_t$, the fraction of mass increase at time t, obeys the relationship

$$Q_t = k\sqrt{t} + \gamma \tag{1}$$

where k, γ are coefficients ($\gamma \neq 0$)

Table 7. Effect of temperature on the rate constants[a]

Set	t, °C	k x 100[b]	P, cm Hg	α' x 100[c]
A1	25.0	12.72	1.430	9.6
A2	30.0	11.68	1.260	10.0
A3	38.5	11.63	1.450	8.6
A4	47.2	9.09	1.375	7.3
B1	22.9	7.47	1.273	6.6
B2	28.2	13.86	1.354	10.9
B3	39.4	10.93	1.354	8.8
B4	43.2	8.28	1.273	7.3

[a] Reproduced from Ref.[11] by permission of the Academic Press Inc.
[b] The unit for k is (g/g dry zeolite) $(min)^{-\frac{1}{2}}$.
[c] Calculated by taking b' = −0.9786 x 10^{-2} (g/g dry zeolite $(min)^{-\frac{1}{2}}$ the unit for α' is (g/g dry zeolite) $(cm\ Hg)^{-1}$ $(min)^{-\frac{1}{2}}$.

Table 8. Effect of temperature on the slope of the second line[a]

Set	t, °C	k_1 x 100[b]	P cm Hg	(k_1/P) x 100
A1	25.0	2.89	1.430	2.023
A2	30.0	2.29	1.260	1.818
A3	38.5	1.59	1.450	1.098
A4	47.2	2.02	1.375	1.474

[a] Reproduced from Ref.[11] by permission of the Academic Press Inc.
[b] The second line refers to the line after the first break in the plots of Fig. 3. The unit for k_1 is (g/g dry zeolite) $(min)^{-\frac{1}{2}}$.

After a certain period of time and for larger uptakes a second equation $Q_t = k_1 \sqrt{t} + \gamma_1$ appears to be followed. The coefficient k correlates linearly with the pressure of monomer P as k = a'P + b' where, a', b' are the coefficients.

The influence of temperature is shown in Tables 7 and 8. The value of the constant a' tends to show a maximum value in the range of 28–30 °C where as k_1/P goes to minimum near 39 °C.

The effects of sorbed water upon the kinetics are shown in Figures 3 and 4. As the amount of zeolitic water increases a maximum value of the rate constant is realised around 4–6 wt. % of zeolite.

The polymerization reaction is partially diffusion controlled. As soon as the first polymer is formed the surface of each catalyst crystal becomes coated with a film of the polymer which increases in thickness as the reaction proceeds. The monomer reaches the active centres on the zeolite surface or within the zeolite by dissolving in the polymer and diffusing through it to the catalyst. The relationship

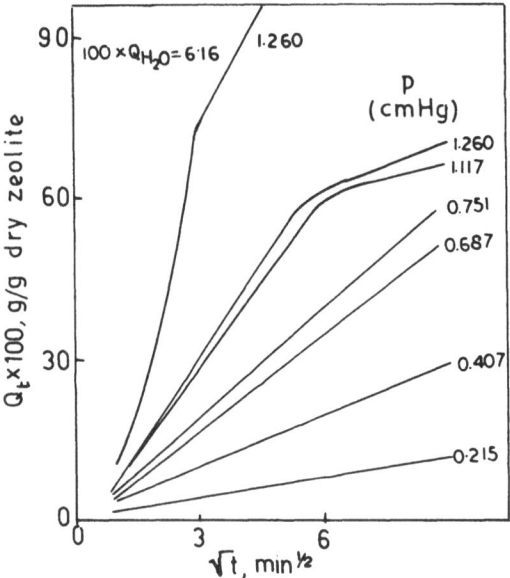

Fig. 3. Mass increase due to monomer uptake, Q_t vs $\sqrt{t}$ plots at several constant vapour pressures of monomers (indicated at top right of each curve); top left: kinetic curve for a zeolite containing 6.16% water[11] by kind permission from Academic Press Inc.

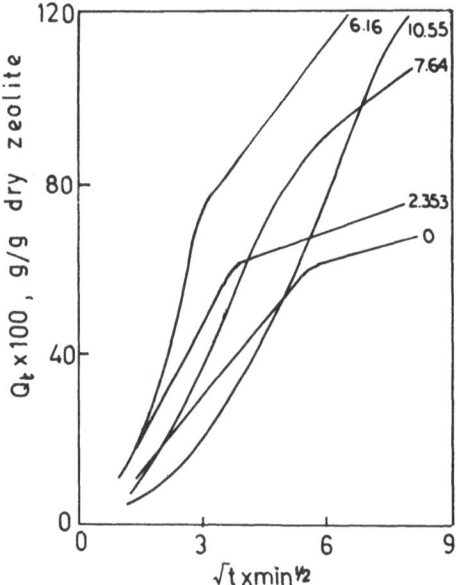

Fig. 4. Q_t vs $\sqrt{t}$ plots obtained with H-mordenite containing different amounts of water (figures at top right of each curve indicating the water content of the zeolite)[11]

between one dimensional diffusion and the amount of polymer formed is of the form

$$Q_t = AX(t) = 2 A \alpha (Dt)^{\frac{1}{2}} \tag{2}$$

where A is the area of the film and α is defined by

$$C_o = \Lambda^{\frac{1}{2}} \alpha \exp \alpha^2 \text{ erf } \alpha$$

assuming the boundary conditions for the monomer to be

$C = C_o$ at x = 0 for all t
$C = 0$ at x = X(t) for t > 0

where x = 0 is the interface between the vapour and polymer and x = X(t) is the interface between the polymer and zeolite crystals. The breaks in the plots of Q_t vs. t in Fig. 3 may be rationalized by supposing a change in the physical configuration of the polymer chain once the amount of the film has increased sufficiently. At this stage the tacky films possibly cohere under surface tension forces and reduce the area A through which monomer diffuses to the catalyst surfaces. Subject to certain limitations the equ.(1) provides a physical basis for observed linear plot of Q_t vs $\sqrt{t}$ and hence of pressure.

Effects of Temperature: The influence of temperature on the present system is rather complex (Tables 7 and 8). In diffusion controlled mechanism, the temperature coefficients of D and α in Eq. (2) should determine the behaviour. D increases with temperature but α decreaes at constant pressure because solution of the monomer vapour in the polymer is exothermal.

Co-catalysis by Zeolite Water. The initial catalytic effect of water (Fig. 5) declines at about *ca.*, 5% by weight (g/g of dry zeolite). This behaviour implies that polymerization on the water free catalyst occurs both at the polymer-catalyst interface and at sites within the crystal. Intracrystalline sites are less accessible and consequently less effective than those at the surface. With increase in the amount of

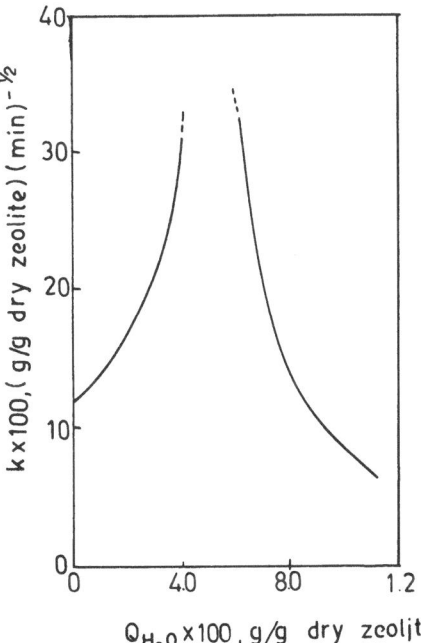

Fig. 5. Rate constant k as a function of the water content of the zeolite[11]

zeolite water, the intracrystalline channels are liable to become more and more blocked and hence still less accessible, so that overall rate of polymerization decreases. Also, excess water molecules may, by preferential adsorption at surface sites, block access of monomer to the active centres on the surface.

Mechanism of Catalysis. H-mordenite is believed to be rich in adjacent Lewis and Brønsted acid sites. Accordingly, a proton transfer from the Brønsted acid sites to the monomer occurs to give carbonium ions, followed by polymerization:

(i) $\quad$ [Si structure with H] $\quad$ [Al structure] $\quad + \quad$ $\overset{R}{\underset{}{CH}}=CH_2 \quad ----\rightarrow$

$\quad$ [Si structure] $\quad$ [Al$^{(-)}$ structure] $\quad + \quad$ $CH_3\overset{(+)}{—}CH \quad$
$\quad\quad\quad\quad\quad\quad\quad\quad\quad\quad\quad\quad\quad\quad\quad\quad\quad\quad\quad\quad\quad\quad\quad\underset{R}{|}$

(ii) $\quad$ $CH_3—\overset{+}{C}H + CH_2=CH \quad ------\rightarrow \quad CH_3—CH—CH_2—\overset{+}{C}H$
$\quad\quad\quad\quad\underset{R}{|}\quad\quad\quad\underset{R}{|}\quad\quad\quad\quad\quad\quad\quad\quad\quad\quad\underset{R}{|}\quad\quad\quad\underset{R}{|}$

(iii) $\quad$ $CH_3—CH—CH_2—\overset{+}{C}H + CH_2=CH \quad ----\rightarrow \quad CH_3—CH—CH_2—CH—CH_2—\overset{+}{C}H—$
$\quad\quad\quad\quad\quad\underset{R}{|}\quad\quad\quad\quad\underset{R}{|}\quad\quad\quad\underset{R}{|}\quad\quad\quad\quad\quad\quad\quad\quad\quad\underset{R}{|}\quad\quad\quad\quad\underset{R}{|}\quad\quad\quad\underset{R}{|}$

$$\ldots\ldots\text{etc.}$$

where R = n-butoxide.

The zeolitic water in presence of the Brønsted sites may also yield some hydronium ions by proton transfer.

(iv)

[Si structure with H] $\quad$ [Al structure]

$+$

H_2O

$\downarrow$

[Si structure] $\quad$ [Al$^{(-)}$ structure] $\quad + \quad H_3O^+$

The hydronium ions can yield by proton transfer carbonium ions of the kind seen in reaction (i), followed by reactions (ii), (iii), etc. The co-catalytic action of water can be understood if reaction (iv) followed by proton transfer from hydronium ion to the monomer occurs more readily than the direct transfer according to reaction (i).

In a sodium mordenite of unit cell composition $Na_8[Al_8Si_{40}O_{96}]24\,H_2O$ the ideal outgassed hydrogen form should have 1.61×10^{21} Brønsted acid sites per gram of dry zeolite. The maximum reaction rate corresponds with 1.7×10^{21} molecules of water added to the zeolite. This quantity of water is just sufficient to convert the hydrogen mordenite to hydronium mordenite. However, for general validity of the water co-catalysis mechanism it is necessary to prove that maximum reaction rates should appear in other hydrogen zeolite catalysts when the water added is equal to the number of Brønsted acid sites.

The polyvinyl ethers produced under these conditions are tacky meaning very low molecular weight ($DP \cong 10$ for poly n-butyl vinyl ether).

Polymerization behaviour of ethyl vinyl ether and isobutyl vinyl ether[12] on H-mordenite and H-faujasite (H-Y) reveals certain distinctive features in comparison with that of n-butyl vinyl ether on these catalysts.

Kinetics. For the polymerization of the vinyl ethers on H-mordenite, the fractional weight increase, Q_t (g/g of dry zeolite), obeys the same $(t)^{\frac{1}{2}}$ law [Eq. (1)] as observed for n-butyl vinyl ether[11].

With H-Y as catalyst, the vinyl ethers polymerize following a different kinetic law in which Q_t is a linear function of ln t,

$$Q_t = k \ln t + \gamma \tag{3}$$

where k and γ are coefficients.

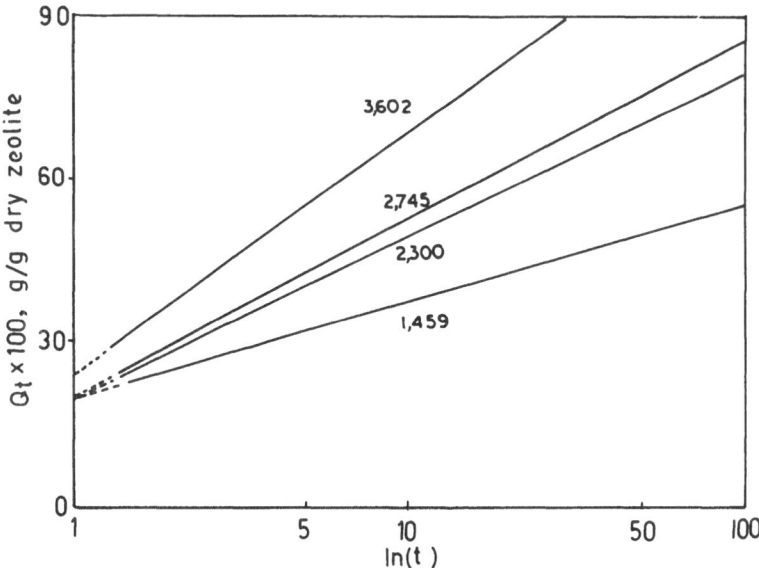

Fig. 6. Q_t plotted against ln t (t in min) at constant vapor pressures of isobutylvinylether indicated on top of each graph[12]

Fig. 6 represents the variation of Q_t vs ln t.

The physical implication of Eq. (3) is that the catalytic centres in H-faujasite (H-Y) are progressively inactivated by polymer growth, possibly as a result of the formation of immobilized polymer clogging the intracrystalline channel of H-Y more and more so that access of monomer to the intracrystalline sites becomes increasingly difficult. However, little or no immobile intracrystalline polymer forms on H-mordenite surface. Consequently surface sites can be active and monomer can always displace polymer segments on an external surface, hence being available at the reaction sites. Further evidence on this interpretation of the kinetics over H-Y as compared with H-mordenite is provided from the observed effect of water on the kinetics and DTG curves.

The Influence of Zeolitic Water. The influence of water uptake on the rate of polymerization of n-butyl ether over H-Y is shown in Figs. 7a and 7b. Small amounts of water increase the slope of the linear part of plots of Q_t vs ln t. For large amounts of water the rates are initially much depressed but subsequently, as in the H-mordenite, show progressive acceleration possibly due to displacement of some water by monomer and polymer as the reaction proceeds. The initial slopes of the curves of Fig. 7a (k values of the equation 3) are plotted against the amount of water, Q_{H_2O} on Fig. 7b. The curve passes through a maximum near 5% uptake of water (g/g of dry zeolite). Thus the small amounts of water functions as a co-catalyst with H-Y as with H-mordenite. The strong variation of the rate of polymerization with water content provides additional evidence that the kinetics are at most only partially controlled by the rate of diffusion of monomer through polymer films growing around each catalyst crystal.

Water Sorption Isotherm. Water sorption isotherm provides additional evidence for the contention that in H-Y, polymerization occurs within the channels of faujasite, in contrast to H-mordenite where little polymer is found in the channels.

Isotherms for water uptake by pure H-mordenite and H-mordenite-polymer-composites with 20.7 and 9.7 wt.% of poly ethyl vinyl ether and poly isobutyl vinyl ether, respectively are shown in Fig. 8a. Water uptake in the polymer is expected to be small and so the uptake of water within the mordenite crystals is little affected by the polymer. This implies that the amount of polymer formed and retained within the mordenite crystals must be small, especially for poly isobutyl vinyl ether.

With H-Y faujasite, the water uptake behaviour presents very distinctive features. Sorption isotherms are shown in Fig. 8b for H-Y and H-Y-polymer composites in which wt.% are respectively 24.3, 29.3 and 27.6 with respect to outgassed zeolite for poly ethyl vinyl ether, poly isobutyl vinyl ether and poly n-butyl vinyl ether. Comparison of greatly reduced water uptakes with that in pure H-Y reveals that about 62, 64 and 69% of the intracrystalline pore volume of H-Y is inaccessible to water for the three composites in the order given above.

Thermogravimetric Analyses. Further significant features on these systems are revealed by thermogravimetric analysis of zeolite-polymer composites after removal of the maximum possible amount of polymer by extraction with chloroform. The curves of weight loss, ΔM, relative to the final mass, M_f, are shown, respectively in Figs. 9a and 9b for H-mordenite alone and H-mordenite + polymer and also for H-Y

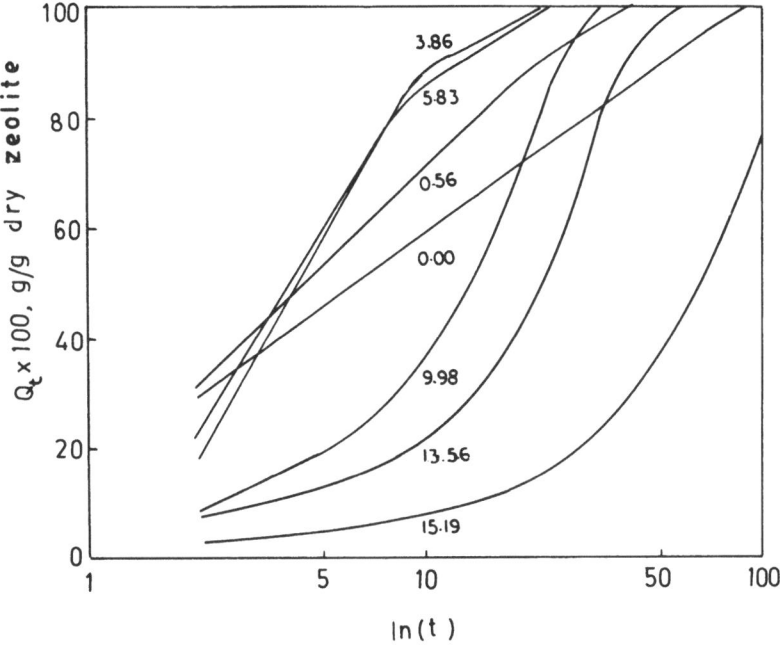

Fig. 7a. Q_t plotted against ln t (t in min) for n-butylvinylether after sorption of water (wt. %) given by the numbers on the curves[12]

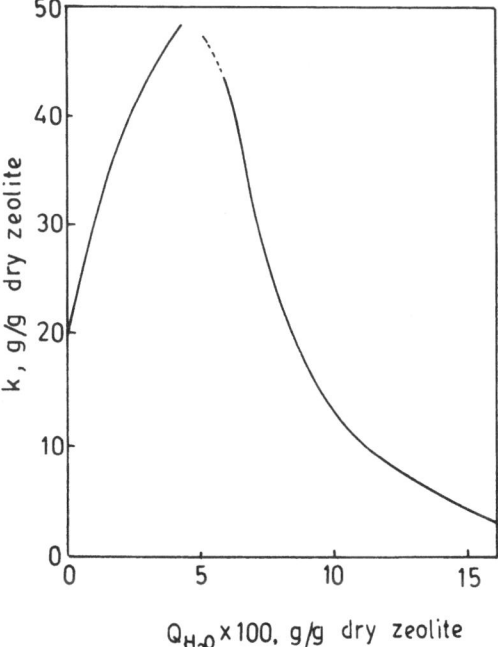

Fig. 7b. k vs Q_{H_2O} plots for n-butyl-vinylether[12]

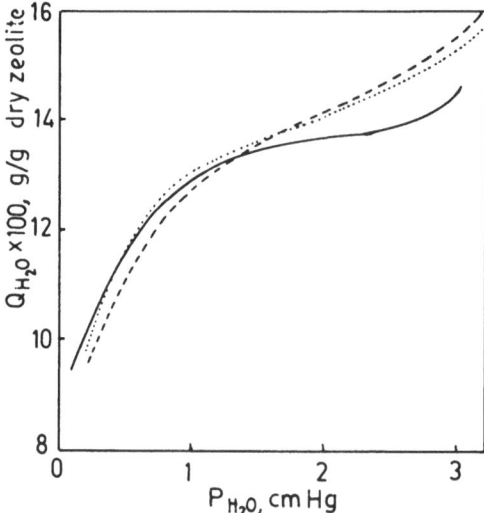

Fig. 8a. Isotherms for sorption of water at 30 °C in pure H-mordenite (—); in H-mordenite +20.7% by wt. of polyethylvinylether (– – – –); in H-mordenite +9.7% by wt of polyiso-butylvinylether (. . . .)[12]

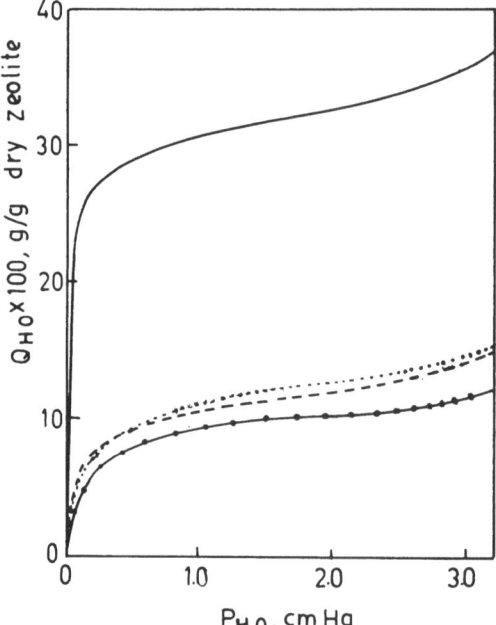

Fig. 8b. Isotherms for sorption of water at 30 °C in pure H–Y (—); in H–Y +24.3% by wt. of polyethyl-vinylether (. . . .); in H–Y +29.3% of polyisobutylvinylether (– – – –) and in H–Y +27.6% poly n-butylvinyl-ether (–•–•–•–)[12]

alone and H-Y + polymer. The amount of polymer not removed by chloroform extraction can be estimated as the limiting weight loss of pure H-mordenite subtracted from those of the mordenite + polymer composites. The amounts so estimated are 1.8, 3.5 and 5% of the weights of dry zeolite for the polymers from isobutyl, *n*-butyl- and ethyl vinyl ether, respectively which possibly corresponds to the upper limits to the quantity of polymer entrained in the intracrystalline channels. The water contents of H-Y and of H-Y + polymer given in Fig. 8b give

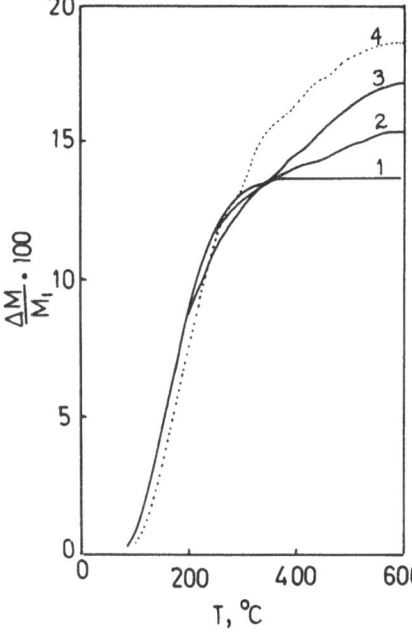

Fig. 9a. Thermogravimetric weight loss (% of final wt. at 600 °C) for pure H-mordenite (1); and composites with polyisobutylvinylether (2); poly n-butylvinylether (3) and polyethylvinylether (4)[12]

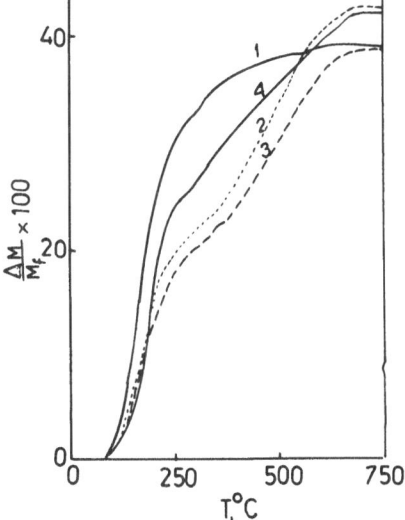

Fig. 9b. Thermogravimetric weight loss curves for pure H–Y (1); H–Y composites with polyethylvinylether (2); polyisobutylvinylether (3) and with poly n-butylvinylether (4)[12]

estimates at 1.4 cmHg of 11.8, 11.0 and 9.5% of water in the dry zeolite for zeolite + polymer (poly ethyl vinyl ether, poly isobutyl vinyl ether, and poly *n*-butyl vinyl ether composites). Calculations on the basis of total weight loss of the composites of Fig. 9b reveal 14, 15 and 16% of unextractable polymer from ethyl,- isobutyl,- and *n*-butyl vinyl ethers respectively. These data therefore clearly suggest that the polymer inclusion complexes form more readily with H-Y than with mordenite. The reason for the specific behaviour may be associated with the sufficiently open

Table 9. Polymerization[a] of N-vinylcarbazole by 13X and SK 500 molecular sieves[b]

Sl. No.	Amount of monomer (N-Vinyl-carbazole)	Amount of zeolite (g)		Polymeri-zation Temp.	Rate $(R_p) \times 10^6$ mol $l^{-1}s^{-1}$
		13X sieves (pellet)	SK 500 (pellet)		
1	0.19	0.20	–	29	–
2	0.097	0.20	–	55	2.90
3	0.097	0.20	–	70	4.89
4[c]	0.097	0.20	–	70	3.92
5[c]	0.097	–	0.20	29	5.5
6[c]	0.097	–	0.20	55	11.0

[a] All polymerizations done in benzene with [N-vinyl carbazole] = 0.1 [M] in air.
[b] Reproduced from Ref.[13] by kind permission of Biswas et al.
[c] Polymerizations done under nitrogen atmosphere.

three-dimensional channel networks compared with crystals having parallel non-intersecting channels and also with the smaller intracrystalline free volume in H-mordenite than in H-Y.

C. Nitrogen Vinyls

Biswas et al.[13] have studied the polymerization of N-vinyl carbazole by 13X and SK 500 molecular sieves under a variety of conditions.

General Features. In general, the polymerizations catalysed by 13X sieves are slow at room temperature usually giving low yields of polymer. Results of Table 9 indicate that above *ca.*, 55 °C with catalyst : monomer ratio, 2 : 1 the polymerization rate increases. Interestingly the polymerization rate is found to be lower in presence of nitrogen gas compared to that in air.

For the polymerization catalysed by SK 500 pellets also the rate of polymerization is slow at ambient temperature, increasing at higher temperature (Table 9).

The polymerizations are exothermic and not accompanied by any typical colour development suggestive of charge transfer polymerization. On the contrary, the SK 500 sieve catalyst undergoes a colour change from white to black in the case of solid pellets and white to green in the case of powdered sieves.

Dependence of Conversion on Concentration, Physical State and Preconditioning of the Molecular Sieves. The polymerizations are significantly dpendent on the concentration, physical state and pretreatment of the molecular sieves. Some typical results demonstrating the effect of variation in the weights of catalyst on the rate of polymerization are indicated in Table 10. Figure 10 represents the typical plots of R_p versus (weight of zeolite) for the polymerization of NVC by 13X and SK 500 pellets respectively. Significantly, Barson et al.[3] reported an order of 0.55 ± 0.11 for styrene polymerization by 13X powdered sieves.

Table 10. Dependence of rate of polymerization[a] on zeolite concentration[b]

Sl. No.	Amount of zeolite (g)		Rate $(R_p) \times 10^6$ mol $l^{-1}s^{-1}$
	13X sieve (pellet)	SK 500 sieve (pellet)	
1	0.20		4.89
2	0.40		5.65
3	0.60		8.39
4		0.097	6.7
5		0.20	11.0
6		0.30	20.0
7		0.38	26.6
8		0.48	33.0

[a] Each polymerization carried out using monomer (N-vinyl carbazole) 0.097 g at 55 °C in nitrogen atmosphere in benzene medium.

[b] Ref.[13] by kind permission of Biswas et al.

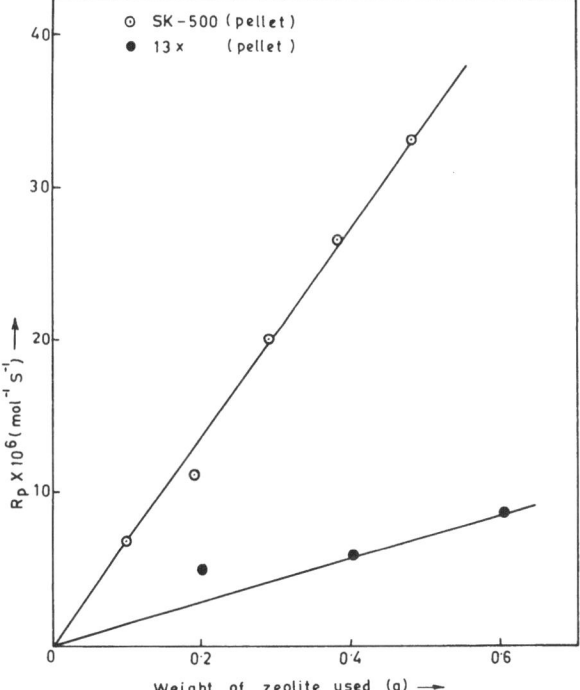

Fig. 10. Dependence of the rate of polymerization (R_p) on the weight of the zeolite; monomer: N-vinylcarbazole (Biswas et al.[13])

The effect of preheating the catalyst at three different temperature ranges on the conversion is illustrated in Table 11. It is evident from these data that the initial rate of polymerization is increased more than one and half times as the preheating temperature of the catalyst is raised from 90 °C–100 °C to 180 °C–190 °C.

Table 11. Dependence of the rate of polymerization[a] on the preheating temperature of the zeolite[b]

Sl. No.	Temperature of the catalyst (13X sieve) pretreatment $°C$	Amount of zeolite (g) 13X sieve (pellet)	Rate $(R_p) \times 10^6$ mol $l^{-1} s^{-1}$
1	90–100	0.20	4.89
2	180–190	0.20	8.10
3	240–250	0.20	8.14

[a] Polymerization done in aerial atmosphere at an average temperature 70 $°C$ in benzene medium using [N-vinyl carbazole] = 0.1 [M].
[b] Reproduced from Ref.[13] by permission of Biswas et al.

Table 12. Dependence of rate of polymerization on physical state of the zeolite[a]

Sl. No.	Amount of zeolite (g)		Physical state of the zeolite	Rate $(R_p) \times 10^6$ mol $l^{-1} s^{-1}$
	13X	SK 500		
1[b]	0.20		Pellet[d]	4.9
2[b]	0.20		Powder[e]	15.5
3[c]		0.097	Pellet[d]	6.7
4[c]		0.097	Powder[e]	11.0

[a] Reproduced from Ref.[13] by kind permission of Biswas et al.
[b] Polymerizations carried out in aerial atmosphere at an average temperature 70 $°C$. [N-vinyl carbazole] = 0.1 [M] in benzene.
[c] Polymerizations carried out under nitrogen at an average temperature 55 $°C$, [N-vinyl carbazole] = 0.1 [M] in benzene.
[d] 1.5 mm in diameter.
[e] 100 mesh size.

Table 13. Dependence of rate of polymerization[a] on N-vinyl carbazole concentration. (Reproduced from Ref.[13] by kind permission of Biswas et al.)

Sl. No.	Amount of zeolite (g) (13X pellet)	Amount of N-vinyl carbazole (g)	$R_p \times 10^6$ mol $l^{-1} s^{-1}$
1	0.20	0.097	4.89
2	0.20	0.1931	15.91
3	0.20	0.3105	29.41

[a] Polymerization carried out in aerial atmosphere at 70 $°C$ in benzene medium.

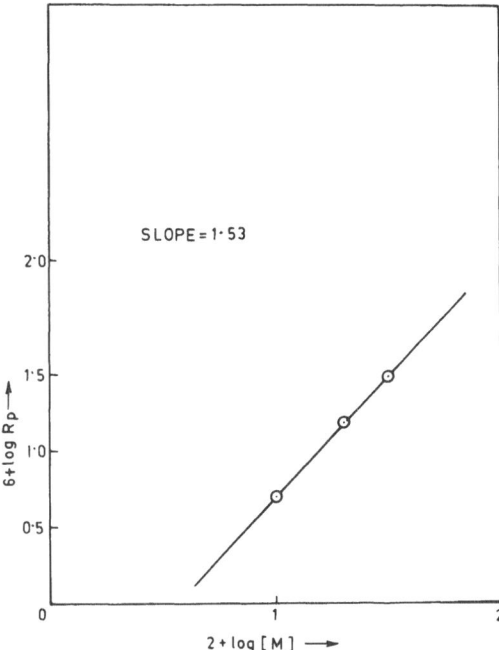

Fig. 11. log Rate vs log [M] plots for the polymerization of N-vinyl-carbazole by 13 X molecular sieve. (Biswas et al.[13])

Further increase in the preheating temperature does not have any marked effect on the rate. Further, the preheating temperature appears to have no effect on the limiting yield of polymerization which is about 40% in three hours with the particular recipe shown in Table 11. It is significant to recall in this context the observation of Barson et al.[3] that for the polymerization of styrene by 13X zeolite the rate of monomer consumption decreases with increase in the preheating temperature. In the present case the enhancement of rate at higher preheating temperatures may be ascribed to the elimination of adsorbed water molecules from the sieve, the polymerization being known[3] to be sensitive to the presence of water.

As is usual with all heterogeneous surface reactions the physical state of the sieve has a significant effect on the catalytic efficiency. The results of several comparative experiments involving pellet and powdered (100 mesh size) sieves are reproduced in Table 12 which amply demonstrate that powdered sieves are more efficient polymerization catalysts than pelleted ones. This is consistent with the fact that with the powdered sieve a greater surface area is available for occupation by the reacting monomeric moieties.

Dependence of Conversion on Monomer Concentration. At a fixed concentration of 13X pellet zeolite, the rate of polymerization increases with increase in N-vinyl carbazole concentration. This is borne out by the results of Table 13. The usual log R_p vs log [M] plot (Fig. 11) indicates a monomer exponent of 1.53. This is characteristically distinct from what is observed in the conventional cationic polymerization of this monomer by aprotonic acid catalyst where the order is consistently unity at low monomer range there after decreasing with further increase of the monomer concentration.

Table 14. Effect of additives[a] on the rate of polymerization[b]

Sl. No.	Amounts of additives	Rate of polymerization $(R_p) \times 10^6$ mol $l^{-1}s^{-1}$
1	–	4.89
2	H_2O (0.05 g)	3.05
3	H_2O (0.1 g)	–[c]
4	HCl (10^{-3}M)	9.78

[a] Reproduced from Ref.[13] by kind permission of Biswas et al.
[b] Polymerization, carried out with 0.20 g of 13X (pellet) zeolite and
    0.097 g of N-vinyl carbazole in aerial atmosphere at 70 °C in benzene medium.
[c] 4% yield after 3 hours.

Table 15. Dependence of degree of polymerization of zeolite concentration[a]

Sl. No.	Types of zeolite	Amount of zeolite (g)		Degree of polymerization[b] $\bar{P}_n$
		Pellet	Powder (100 mesh size)	
1	13X	0.40	–	10.1
2	13X	0.60	–	8.5
3[c]	13X	0.20	–	17.8
4[c]	13X	0.20	–	26.5
5	SK 500	0.097	–	8.0
6	SK 500	0.20	–	8.3
7	SK 500	0.29	–	9.3
8	SK 500	0.38	–	11.2
9	SK 500	0.48	–	12.4
10	SK 500	–	0.039	19.7
11	SK 500	–	0.064	20.3

[a] Reproduced from Ref.[13] by kind permission of Biswas et al.
[b] All polymerizations conducted in benzene at 70 °C for runs 1–4, at 55 °C for runs 5 – 9
    and at 29 °C for runs 10 and 11. Molecular weights determined cryoscopially in benzene.
[c] [N-vinyl carbazole] = 0.2 mol l^{-1} in run 3 and 0.32 mol l^{-1} in run 4 and 0.1 mol l^{-1} in
    rest of the experiments.

It is relevant to recall in this context Barson's results[3] which indicate a
monomer order of 1.47 ± 0.19 for the polymerization of styrene in the presence of
13X sieves.

It is felt however that the experimental data are not adequate for the interpretation of the monomer dependence observed in the present context. Further
investigations on the monomer dependence of the rate of N-vinyl carbazole polymerization by SK 500 are in progress[13].

Effect of Additives on the Rate of Polymerization. The conventional cationic
nature of the sieve catalyzed polymerization of N-vinyl carbazole is endorsed by the

observed effect of water, protonic acids, and amines on the rate (Table 14). The retarding action of water and the inhibitory action of amine on such polymerization are reminiscent of what is usually observed in the cationic polymerization of this monomer by protonic acids. The enhancement of the rate with added protonic acids is consistent with the fact that additional protons will be likely to increase the rate of the initiation reaction already occurring in the presence of the sieves.

Dependence of Degree of Polymerization on Zeolite Concentration. Data presented in Table 15 reveal that degree of polymerization for these sieve catalysed polymerization reactions tends to increase slightly with zeolite concentration and appreciably with monomer concentration. Apparently these features are characteristically different from the usual behaviour of degree of polymerization in the polymerization of this monomer by aprotonic acid where the $\bar{P}_n$ values are consistently independent of catalyst and monomer concentration respectively[14]. As pointed out earlier, Barson et al.[3] suggest that for styrene polymerization, molecular weight should not depend upon the weight of zeolite. However a slight increase in the $\bar{P}_n$ value with increase in monomer concentration may be expected due to an increase in the rate of propagation relative to the rate of termination. For the polymerization of N-vinyl carbazole by molecular sieves the observed tendency of the $\bar{P}_n$ to increase slightly with the weight of the zeolite is apparently contradictory.

Relative Efficiencies of SK 500 and 13X as Polymerization Catalysts. For the polymerization of N-vinyl carbazole, the zeolite catalysts in the powder form are more active than the same in pellet form. This is to be expected since a greater effective surface area is available with the powder. Further under all other identical experimental conditions SK 500 sieves appear to be more active than 13X types.

D. Miscellaneous Monomers

Ethylene. The highly reactive H-Y and rare earth X-zeolites have been reported to polymerize ethylene[15, 16, 17], propene[15], hex-1-ene[16] and 2,3-dimethylbut-1-ene[18]. Venuto et al.[16] use acidic faujasites rare earth exchanged X and Y and Hydrogen Y as catalyst. The rare earth X catalyst was prepared by exchanging Linde 13X catalysts with 5% mixed rare earth chloride ($RECl_3, 6H_2O$) solution until a sodium level of 0.58 % by weight was obtained. A typical analysis result of the rare earth mixture (as % weight oxide) employed in these exchanges is La_2O_3 (24.1), CeO_2(48.0), Pr_6O_{11} (5.9), Nd_2O_3(19.1), Sm_2O_3(2.0), Gd_2O_3(0.7) others (0.2). The rare earth Y catalyst (REY) was prepared similarly by base exchanging a synthetic sodium Y aluminosilicate (NaY) with 5 % $RECl_3, 6H_2O$ solution until a sodium level of 0.99 % wt. was obtained. Ethylene reacts with such type of REX to produce low molecular weight paraffin gases through intracrystalline reaction. Deuterated REX converts hex-l-ene into di and trimers. Alkylation and polymerization reactions on these catalysts are enhanced[16] in the presence of proton donors. For the alkylation of benzene with ethylene an acceleration in rate is observed[16] when the REX catalyst is treated with 1.85 meq/g of water, HCl or HBr immediately prior to reactions. Conversely a decrease in the alkylation activity with REX precalcined at temperatures 400 °C is consistent[16] with proton loss through dehydration sites for alkylation in REX and REY arise through an essentially reversible transfer of a proton from a residual strongly held water molecule to the oxygen of the associated AlO_4 tetrahedron.

$$RE.^{3+}O\!\!\begin{array}{c}{}^{-H}\\{}^{-H}\end{array} \quad O^--Zeol \;\rightleftharpoons\; RE^{2++}\!-\!O\!\!\begin{array}{c}\overset{\delta-}{}\;^{-H}\\ H\cdot\\ \delta+\end{array}\!\!\cdots\!\bar{O}\!-\!Zeol \;\rightleftharpoons$$

$$RE^{2+}\!-\!O\!\!\begin{array}{c}{}^{-H}\\{}\end{array} \qquad H^+ \;\; \bar{O}\!-\!Zeol$$

A model for the acid-faujasite catalysed process is suggested[16] through the reaction of 1-hexene and benzene at 80 °C over deuterated REX in the light of the above mechanism.

$$\begin{array}{c}//\ \underline{O}\!-\!Zeol\ //\\ D+\\ \vdots\end{array}$$

$$CH_2\!=\!CH\!-\!C_4H_9 + D^+\ \bar{O}\!-\!Zeol \;\rightleftharpoons\; CH_2\!\overset{+}{=}\!CH\!-\!C_4H_9$$

$$\begin{array}{cc}///\ \underline{O}\!-\!Zeol\ /// & ///\ \underline{O}\!-\!Zeol\ ///\\ \overset{+}{} & \overset{+}{}\\ \rightleftharpoons\; CH_2D\!-\!\overset{+}{C}H\!-\!CH_2C_3H_7 \;\rightleftharpoons\; & CH_2D\!-\!CH_2\!-\!\overset{+}{C}H\!-\!CH_2\!-\!C_2H_5\end{array}$$

A similar mechanism may therefore be suggested for ethylene polymerization over such catalyst (REX) surfaces which would proceed essentially through a Rideal-like pattern.

Table 16 illustrates[16] the product distribution with temperature for the reaction of ethylene with REX. These data reveal a complex series of reactions initially involving acid catalysed polymerization of ethylene to low molecular weight aliphatic polymer. Subsequent intermolecular hydrogen transfer and dehydrogenation-cyclization reactions followed by isomerization and cracking reactions account for the formation of the other products.

Cationized zeolites[19], prepared from Linde 13X by ion exchange with salt solution of K, Tl, NH_4, Mg, Ca, Sr, Ba, Zn, Cd, La, Ce, Mn, Co and Ni catalyse on the polymerization of ethylene, propylene and double bond migration of 1-butene. Ca^{+2} and Cr^{+3} forms of Y zeolites induce C_2-C_5 polyolefin formation[20] at higher temperatures and pressure.

Ion exchange introduction[21] of 1 % Pd into CaNaY zeolites and subsequent heat treatment at 380–450°C yields a catalyst suitable for converting ethylene to C_5 and higher hydrocarbons in 83 % yield. Treatment with hydrogen or helium decreases the activity of the catalyst.

Nobuyashi et al.[22] describe the modification of SK40 [NaY type zeolite] with Cr^{+3}, which is effective for the polymerization of ethylene. The catalyst preparation involves keeping 50 gm SK 500 in 100 ml of 0.5 N $CrCl_3$ at 70 °C with replacement every 4 hours of the $CrCl_3$ solution. The zeolite thus treated is sub-

Table 16. Variations in product distribution with temperature for reaction of ethylene with REX[a]

Temperature (°C)	Major products	
	Intracrystalline	Gaseous
93	Liquid aliphatic polymer	Nil
121	Mixed aliphatic and aromatic polymer	Paraffins (traces)
149	Largely aromatics, small amounts aliphatic polymer	Paraffins (small amounts)
177	Exclusively aromatics	Paraffins (large amounts)
213	Exclusively aromatics[b]	Paraffins (large amounts)

a Reproduced from Ref.[16].
b 23 % of alkylbenzenes (C_{15} and C_{16} predominant), at least 5 % alkyl naphthalenes (C_{18}, C_{17}, C_{20} most prominent), with the remainder consisting of a complex mixture (average molecular weight 490 by vapour pressure lowering) of alkylated polycyclic aromatics and their partially dehydrogenated analogues.

sequently washed free of Cl$^-$ and dried for 12 hrs at 110 °C. About 80 % replacement of Na by Cr occurs by this procedure. Catalyst activation is done for 2 hrs in vacuo at 400 °C.

Cr-Y[23] zeolite catalysts are particularly active for the bulk polymerization of ethylene resulting in the formation of polymers (m. p. 134—42°) with crystalline and unbranched chains. The optimum evaluation temperature of catalyst activation is 350 °C and the yield of polyethylene increases linearly with ethylene pressure (range 5—50 atm.) and time. The active sites on the Cr-Y catalyst appear to be composed of Cr^{+2} ions which remain supported on the zeolites.

Tetrafluoroethylene. Polymerization of tetrafluoroethylene[24] adsorbed on various sorbents including silicagel, NaY zeolites and CaA zerolites is initiated by γ-irradiation. The threshold temperature for the polymerization of the adsorbed monomer increases in the order

silicagel < NaY zeolites < CaA zeolites

which is the same order of increase of the bond strength between adsorbed monomer and sorbents. The polymerization is thought to be free radical in nature and sensitized by the sorbents apparently due to retention of high concentration of free radicals on their surface. However, the mobility of the adsorbed molecules is significantly affected by increasing the temperature. Co-polymerization of tetrafluoroethylene on silicagel depends on the amount of adsorbed monomer and not on its concentration in the gas phase, which suggests that chain growth precedes adsorption of tetrafluoroethylene molecules.

Propylene. Norton[25, 26] first observed olefin polymerization over synthetic molecular sieves. Varieties of type A(3A, 4A, 5A, 5P) and two of type X(10X and 13X) molecular sieves were used to study the polymerization of ethylene, propylene, and isobutylene. Polymerizations were conducted in autoclaves at 200 ° to 500 °C and 100 to 1500 p. s. i. g.

The polymerization rates are found to be first-order in the presence of olefin:

$$\text{Rate of polymerization} = \frac{-d(P_{olefin})}{dt} = -k(P_{olefin})$$

The first-order rate constants (Table 17) computed from the pressure versus time plots are obviously dependent only on molecular sieve types and compositions.

The order of the sieve activities with respect to the polymerization of these monomers is

10X > 13X > 5A > 5P > 4A > 3A = 0

and the order of olefin reactivities over sieves is

isobutylene > propylene > ethylene.

For the type A sieves the order of catalytic activity (3A < 4A < 5A) is virtually the same order as the increasing solid acidity titres (acidity of the solid sieves) and implies that acidity may be affected by the nature of the exchangeable alkali ions. The differences in the polymerization activities are also exposed in Fig. 12 where log k values are plotted against the extents of sieve ion exchange of the sodium form towards the calcium form. With both the A and X type sieves, the calcium form is more reactive than the sodium form. The results further suggest that two types of sieves form two distinct classes whose catalytic activities are dependent on their inherent silica-to-alumina ratios and structures, as well as their alkali and alkaline-earth ion contents. The importance of the silica-alumina ratio in the sieve compositions is underscored by the fact that the log k values correlate inversely with percent reactive alumina contents of all the sieves possessing solid acidity titres, with or without clay binder (Fig. 13). The more catalytic and reactive type X sieves have the lesser proportions of reactive alumina. This behaviour is consistent with the fact that olefin polymerization is initiated by a proton addition process

Table 17. Polymerization of propene over synthetic molecular sieves[a]

Commercial designation, Nominal sieve pore size, A	Composition	Propene polymerization rate constant, $k \times 10^4$, hr^{-1}	Solid-acidity Meq/gx10^2	Relative alumina wt.%
3A	$K_2O.Al_2O_3.2SiO_2.xH_2O$	0.00	−8.3	12.15
4A	$Na_2O.Al_2O_3.2SiO_2.xH_2O$	8.45	3.5	10.9
5A	$1/3\ Na_2O.2/3\ CaO.Al_2O_3.$ $2SiO_2.xH_2O$	108.00	4.5	6.60
5P[b]	$1/3\ Na_2O.2/3\ CaO.Al_2O_3.$ $2SiO_2.xH_2O$	23.2	5.9	8.81
10X	$1/4\ Na_2O.3/4\ CaO.Al_2O_3.$ $2.8SiO_2.xH_2O$	468.00	9.7	3.64
13X	$Na_2O.Al_2O_3.2.8SiO_2.xH_2O$	193.00	19.5	4.17

[a] Reproduced from Ref.[26] by permission of the American Chemical Society.
[b] Commercial sample of powdered 5A sieves without clay binder.

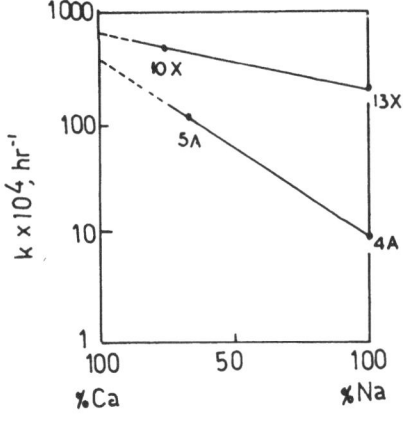

Fig. 12. Dependence of propene polymerization rate constants on molecular sieve types and compositions (Ref.[26] by permission of the American Chemical Society)

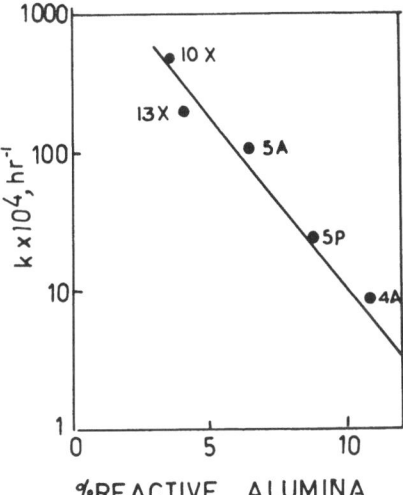

Fig. 13. Correlation of propene polymerization rate constants with reactive alumina contents of molecular sieves[26]

and reactive alumina sites being readily susceptible to hydrolysis producing exchangeable hydroxide ions would be antithetical to acid-catalysis.

Mechanism. Analysis of the lower boiling hydrocarbon fraction from the liquid propene polymer indicates the presence of 2-methyl-l-pentene, 2-methyl-2-pentene, 4-methyl-1-pentene, and cis- and trans-4 methyl-2-pentene. These reaction products are those anticipated for a typical acid catalyzed polymerization of propene as outlined below.

The presence of branched-chain olefin hydrocarbons in the conjunct polymer suggests that the major part, if not all, of the reaction with the 3A, 5A, and 5P molecular sieves must have taken place at acid sites outside of the pores because the branched-chain hydrocarbon molecules produced can not pass out of the small pores of these molecular sieves. In the case of larger pore size 10X and 13X molecular sieve sample, part of the action may take place at external acid sites and part at acid sites within the pores, as the larger pore size of these latter molecular sieves can permit the absorption and desorption of branched-chain hydrocarbons.

2−methyl−2−pentene   2−methyl−1−pentene

4−methyl−1−pentene   trans−4−methyl   cis−4−methyl
                        −2−pentene   −2−pentene

2−methyl−2−pentene   trans−4−methyl   cis−4−methyl
                        −2−pentene   −2−pentene

Propylene polymerization is induced[27] by a suitable catalyst prepared by refluxing 13X molecular sieves in aqueous $Ni(NO_3)_2$, washing in distilled water and drying followed by activation in N_2 at 550 °C. In a typical reaction 12 g of this catalyst yield 322 g of polymer with 1500 p. s. i. g. in an autoclave at 150 °C. The percentage distribution of the products is 68 % C_6, 19 % C_9 and 13 % C_9 fraction.

The catalytic activity of NaX and CaX types zeolite towards propylene polymerization appears to be increased with substitution of Na^+ and Ca^{+2} ions by Co^{+2} ions[28]. However the activity of zeolite does not change till 30 % of Ca^{+2} ions are substituted by Co^{+2} ions. Interestingly, prior γ-ray irradiation of CaX and Co CaX (30.2 % and 53.5 % Co^{+2}) zeolite, increase their catalytic activity. The NaX zeolite activity is however unaffected by this substitution. The polymerization rate is increased with increase in the reaction temperature and preirradiation of zeolites. NiX and CoX zeolites[29] have also been used for the oligomerization of propylene. NiX at 190 °C yields 96 % propylene oligomers while CoX zeolites produce a mixture of paraffins and olefins under identical conditions. The CoY[29] and NaY zeolites, however, yield C_{4-9} paraffins at 190 °C. CaNaX and NdNaX[30] zeolites are superior catalysts compared to synthetic amorphous alumino silicates towards propylene oligomerization. Oxides of Fe, Co, Ni do not in particular modify the catalyst in respect of general activity and isomerization. Preheating of NdNaX zeolites at 300–700 °C inhibits the isomerization reactions. For the sieve catalyst the major oligomerization products include olefin and paraffin of C_{2-8} com-

position, while the amorphous alumino silicates yield predominantly C_{4-8} hydrocarbons.

Lafer et al.[31] suggest that the polymerization of propylene on H-form of
mordenite, erionite and Y type zeolites proceeds by an acid base mechanism involving Brønsted acidic OH groups. For the mordenite type zeolite the hydro-
xonium groups are responsible for the reaction. De-aluminization of H-form
mordenite (NaM) inhibits the polymerization, an observation which, however,
contradicts Norton[26].

1-Butene. Acidic alumina silicate zeolite catalysts[32] with alumina 13 % of
surface area 200 m^2/g activated at 550 °C have been used for isomerization and
oligomerization of 1-butene. The oligomers (37 %) include octene47 %, dodecene
47 %, tetradecene-hexdecene 18 % and eicocene 5 %. 90 % of the residual butene
isomerizes to 2-butene in course of 25 minutes. NaHY zeolites[33] also have been
reported to oligomerize 1-butene at room temperature. Saturated products are
supposed to result from the interaction of OH groups on the zeolite surface with
1-butene. This conclusion is endorsed by the fact that there is no reaction at 125 °C
of 1-butene on dehydroxylated surface of NaHY obtained by treatment with
formic acid and evacuated at 500 °C while 1-butene adsorbed by NaY zeolites (with
OH-groups) slowly oligomerizes at 120 °C.

Isobutylene. Barrer[34] first reported observation of the polymerization of
butylene isomers to a liquid product over the natural molecular sieve, chabazite. In
a later work, Barrer et al.[35, 36] reported that t-butyl chloride and butyl bromide
are decomposed on the external surface of chabazite to give corresponding halogen
acids and isobutylene the latter monomer polymerizing to a nonvolatile oil.

$$CH_3-\underset{\underset{CH_3}{|}}{\overset{\overset{CH_3}{|}}{C}}-Cl \quad ----\blacktriangleright \quad \underset{\underset{CH_3}{/}}{\overset{\overset{CH_3}{\backslash}}{C}}=CH_2 + HCl$$

5A zeolite molecuar sieves[37] have been found to yield a highly branched
product comprising 50 % trimer, some dimer and traces of tetramer from iso
butylene vapour at 15–200 °C and atmospheric pressure. The polymerization catalytic behaviour of 5A sieves is inhibited[38] by neutralizing the acid sites with an
organic N-base whose size is too large to permit entry into the pores of the sieves.
Pretreatment[39] of 5A sieves with excess isobutylene or di-isobutylene at
200–500 °F for 0.5–2.0 hrs followed by stripping with N_2 at 400–800 °F for
2–8 hrs and subsequent cooling to 60–130 °F causes polymerization and C-depo-
sition to occur on the exterior surfaces but not within the pore of the adsorbent.
Isobutylene containing 0.4 % by weight of n-butylene can subsequently be rendered
free of n-butylene on being brought into contact with the 5A sieves preheated as
above.

Norton[26] reported that isobutylene polymerizes more readily than propylene
or ethylene on various molecular sieves of X and A type[26]. In the absence of any
definite characterization of these products, Rhein et al.[40] examined a variety of
molecular sieves under varied conditions of the reaction temperature and time for
isobutylene polymerization. The various molecular sieves used are: type 3A, 3Å;

Table 18. Polymerization[a] of isobutylene on various molecular sieves[b]

Molecular sieve type and size	Re-action time (days)	Stirring	Unreacted iso-butylene	Yield of polymer %	Number average molecular weight $\overline{M}_n$	Weight average molecular weight $\overline{M}_w$
3A, 1/16 pellets	7	No	100	0		
4A, 100/200 mesh	7	No	100	0		
5A, 1/16 pellets	10	No	0	22.7	555	1325
5A, 40/50 mesh	4	Yes	0	54.3	350	–
5A, 40/50 mesh	4	Yes	0	52.6	322	1032
5A, 40/50 mesh	10	No	0	18.9	515	1264
5A, 80/90 mesh	10	No	0	43.6	469	1397
5A, 120/130 mesh	4	Yes	0	56.3	526	1641
5A, 150/170 mesh	10	No	0	48.7	663	1963
5A, 160/170 mesh	1	Yes	0	9.6	1241	3669
5A, 160/170 mesh	1	Yes	0	8.0	1229	3438
5A, powder, 1 – 4 μm	10	No	0	0	–	–
10X, 100/120 mesh	7	No	0	4.5	300	480
10X, 100/120 mesh	9	Yes	0	3.9	575	1390
13X, 100/110 mesh	7	No	50	7.8	2569	5989
13X, 100/110 mesh	9	Yes	13.3	10.9	1140	4896
SK 45 powder, 1 – 4 μm	5	Yes	–	–	839	2566
SK 45 powder, 1 – 4 μm	9	Yes	76.0	0.2	1417	3292
SK 100, 1/16 pellets	5	Yes	–	–	315	365
SK 100, 1/16 pellets	9	Yes	0	5.6	639	1830
SK 110, 1/16 pellets	3	Yes	–	–	982	2164
SK 110, 1/16 pellets	9	Yes	0	4.4	832	1950
SK 200, 1/16 pellets	3	Yes	–	–	781	1232
SK 200, 1/16 pellets	9	Yes	0	0.5	554	1124
SK 310, 1/16 pellets	3	Yes	0	1.2	457	761
SK 400, 1/16 pellets	3	Yes	60	2.2	448	2114
SK 400, 1/16 pellets	9	Yes	66.6	2.5	1233	3807
SK 410, 1/16 pellets	3	Yes	80	0	–	–
SK 500, 1/16 pellets	3	Yes	0	2.0	610	1603

[a] In these polymerizations, 75 ml of isobutylene added to 0.025 kg of molecular sieve (pretreated by heating to 373–473 K and evacuating to 0.133 N/m^2 for 1 day) in a 250 ml bulb.
[b] Reproduced from Ref.[40] by permission of Polymer.

type 4A, 4.2 Å; type 5A, 5 Å; type 10X, 8 Å and type 13X, 10 Å . SK 45 (a potassium-exchanged type L)[41, 42]; SK 100 (decationized type Y[43, 44] with 0.5 % Pd[45]); SK 110 (partly decationized type Y with 5.2 % MnO and 0.5 % Pd[45]); SK 200 (calcium-exchanged type Y with 0.5 % Pt[45]); SK 310 (calcium-exchanged type Y with 0.5 % Pd[45]); SK 400 (sodium-exchanged type Y with 1.0 % Ni[45]); SK 410 (sodium-exchanged type Y with 1 % Cu[45]); and SK 500 (type Y with 35 % cations 15 % Na_2O and 50 % mixed rare earth oxides[46]). Salient experimental conditions and results are presented in Table 18.

Results of detailed investigations on the polymerization of isobutylene by 13X sieve are collected in Table 19.

Analysis of these experimental data along with the character of molecular sieves reveals some interesting relationships between the dependent variable to the independent variables in these polymerizations (Table 20).

The polymerization of isobutylene by molecular sieves occurs at cationic sites within the molecular sieve cavity. This is supported by the fact that

(i) polymerization only occurs when the cavity pores are $\geqslant 5 \times 10^{-10}$ m ($\geqslant 5$ Å)
     [i. e. the isobutylene cannot enter through pores $< 5 \times 10^{-10}$ m (< 5 Å)]
(ii) polymer yield and fraction of isobutylene reacted are greatest for molecular sieves with cationic exchanged sites
(iii) polymer is essentially monofunctional in unsaturation.

The polymerization reaction is rather slow, since both polymer yield and fraction of reacted isobutylene increases over a period of days. In addition, the polymer molecular weight decreases with increasing temperature, as generally observed in isobutylene polymerization.

Table 19. Processing conditions and experimental results for the polymerization of Iso-butylene by molecular sieve 13X (Reproduced from Ref.[40] by permission of Polymer)

Exp. No.	Reaction time (days)	Temp. (K)	Ratio of catalyst to monomer (g/ml)	Stirring	Un-reacted iso-butylene (%)	Yield of polymer	Number average mol. weight $\overline{M}_n$	Weight average mol. weight $\overline{M}_w$
1	0.333	298	0.333	Yes	66.6	2.6	911	3275
2	1	298	0.333	Yes	57.8	7.1	1565	4935
3	2	298	0.333	Yes	26.7	4	1293	4405
4	4	298	0.333	Yes	22.6	24.4	802	4279
5	6	298	0.333	Yes	0	14.7	863	4933
6	12	298	0.333	Yes	5.3	11.1	889	3971
7	15	298	0.333	Yes	6.7	16.7	770	4229
8	20	298	0.333	Yes	6.7	16	839	4049
9	9	298	0.5	Yes	19.0	43.4	865	4688
10	9	298	0.25	Yes	4	11.3	807	4135
11	9	298	0.5	Yes	2	30.3	586	3081
12	9	298	1	Yes	0	12.2	1264	3500
13	9	298	2.5	Yes	0	8.3	717	1405
14	1	193	0.333	No	97.6	0.1	1715	5267
15	4	193	0.333	No	31.4	12.2	780	3257
16	7	193	0.333	No	72	4.7	1107	5210
17	1	273	0.333	No	31.4	9.1	1137	5399
18	4	273	0.333	No	25	21.3	712	3369
19	7	273	0.333	No	0	30	820	3363
20	1	298	0.333	No	54.7	9.6	489	2507
21	4	298	0.333	No	20	22.9	613	3095
22	1	373	0.333	No	0.7	11.0	971	5710
23	4	373	0.333	No	6.7	16.0	393	1454
24	7	373	0.333	No	9.3	33.1	370	1678

Table 20. Parameters influencing the characteristics of the reaction products from the molecular-sieve catalysed polymerization of isobutylene (Reproduced from Ref.[40] by permission of Polymer)

Independent variable	Dependent variable		
	Polymer yield (%)	Polymer mol. weight, $\bar{M}_n$	Isobutylene reacted (%)
Pore diameter	No yield $\leqslant 4.2 \times 10^{-10}$ m (4.2 A) maximum at 5×10^{-10} m (5 A)	Inreasing with increasing pore diameter	None $\leqslant 4.2 \times 10^{-10}$ m (4.2 A) maximum $(5.8 \times 10^{-10}$ m $(5-8$ A)
Exchanged cation	Generally cations > cations + MnO > Na$_2$O > (Na$_2$O + cations + rare earth oxides) > CaO	Generally (cations + MnO) ~ Na$_2$O > CaO > (cations + Na$_2$O + rare earth oxides) > cations	Generally (cations) ~ (cations + Na$_2$O) ~ CaO ~ Na$_2$O + cation + rare earth oxides) > Na$_2$O
Type of sieves	A > X > Y > L	L > Y, L > A, other wise indeterminate Pt > Pd	A $\geqslant$ Y $\geqslant$ X > L Pt = Pd, Ni > Cu
Catalyst on sieve particle size	Pd > Pt, Ni > Cu Increasing yield for increased size	Increasing $\bar{M}_n$ for decreasing particle size	Interdeterminate
Temperature	Increased	Decreased	Increased
Time of reaction	Increasing	Generally decreased	Generally increasing
Ratio catalyst/ monomer	Decreased	Decreased	Decreased
Effect of stirring	Decreased	Increased	Increased

More recently[47, 48] it has been observed that exchange of Na$^+$ ions in X and Y zeolites with Ni^{+2}, Cu^{+2}, Nd^{+3}, enhances the acidity and degree of isobutylene polymerization.

2-Alkoxy Propene.[49] Linde 4A, Na-alumino silicate molecular sieves polymerize 2-alkoxy propenes to products with molecular weight as high as 100,000–250,000. Nearly 44 % polymer formation has been reported at −10 °C over 24 hours and increasing the temperature to 60 °C enhances the yield to 72 %.

2,2-Dimethyl-4-methylene-1,3-dioxolane and 2-Methyl-4-methylene-1,3-dioxolane. Goodman and Abe[50] report that the polymerizations of 2,2-dimethyl-4-methylene-1,3-dioxolane and 2-methyl-4-methylene-1,3-dioxolane proceed[50] in presence of 4A and 5A molecular sieves respectively yielding 7 % and 70 % yield of polymer with intrinsic viscosity 0.04. However other conventional cationic catalysts such as BF$_3$.Et$_2$O, AlCl$_3$, etc are more efficient and produce polymers with higher intrinsic viscosities. The polymerizations proceed by a cationic mechanism and take place through a coupled vinyl and acetal ring opening to give polyketoethers.

Cationic catalysts

$CH_2=C\overline{}CH_2$

R_1 R_2

$$\overline{}\left[\begin{array}{c} O \\ \parallel \\ CH_2-C-CH_2-O-\overset{R_1}{\underset{R_2}{C}} \end{array}\right]_n$$

Infrared and ultraviolet absorptions suggest these polymers to possess appreciable carbonyl content but carbonyl absorptions diminish in the following order:

$$BF_3.Et_2O\ (-78\ °C)\quad \geqslant AlCl_3\ (25-30\ °C) > \text{molecular sieve (above 25 °C)}$$
$$\cong\ H_2SO_4\ (\text{concn.})\ (-78\ °C) > AlCl_3\ (-78\ °C)$$

Phenol. An interesting work on the oxidative polymerization of phenol in presence of molecular sieves is revealed in a patent by General Electric Co.[51]. Polyphenylene oxides have been obtained by the oxidation of 2,6-dimethyl phenol in the presence of Cu-amine complex catalyst and 4A molecular sieves in toluene in presence of oxygen, flowing at the rate of 3–4 ft³/hour for 2 hours at 25 °C. Figure 14 illustrates the dependence of the polymer molecular weight on the ratio of molecular sieve to monomer. Regeneration of used up molecular sieves is also possible by conventional method.

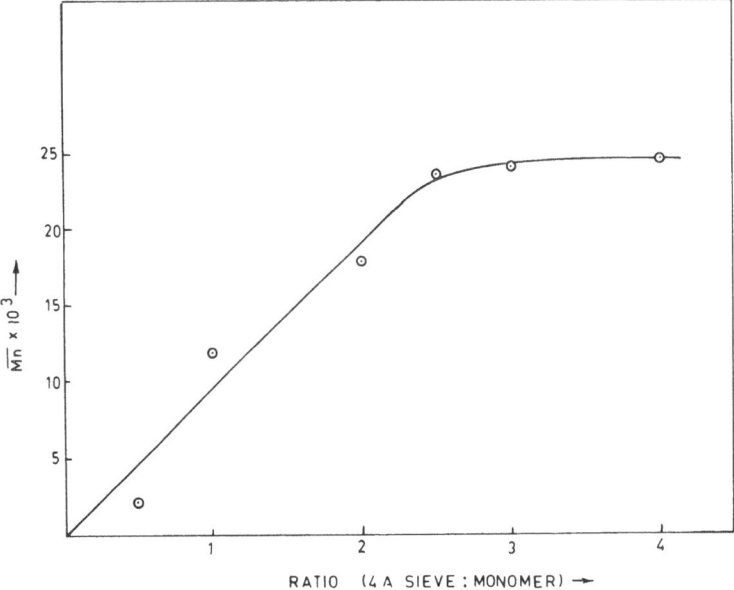

Fig. 14. Dependence of $\overline{M}_n$ of polyphenylene oxides on the ratio of 4 A molecular sieve to monomer (2,6 dimethylphenol) (from the data of Ref.[51])

Isoprene and Acrylonitrile. Pd-zeolites[52] containing a cobalt salt in presence
of carbon monoxide and hydrogen induce the polymerization of monomers such as
– isoprene, styrene, and acrylonitrile. Oil soluble cobalt salts are in general more
effective and cobalt carbonyl formed on the zeolite surface probably serves the role
of the auxiliary catalyst. A mixture of 15 cc acrylonitrile, 0.1 gm Pd-zeolite, and
0.005 g cobalt naphthalate in a 100 cc autoclave charged with 100 atmosphere 1 : 1
carbon monoxide and hydrogen at 115 ° gives 45 % powdered poly acrylonitrile of
average molecular weight 7.9×10^4, the yield increasing to 90 % in presence of
carbon tetrachloride as solvent. Likewise from styrene (15 cc) with 0.1 g Pd-zeolite
and 0.1 g cobalt naphthalate, 40 % clear polystyrene of molecular weight 92×10^4
is formed. Isoprene more or less under identical conditions yields 20 % rubber like
polymer. The mechanisms operative in these cases probably are not cationic in
nature since acrylonitrile is most reluctant to polymerize cationically and cobalt
carbonyls are known to bring about free radical polymerization of vinyl
monomers[53].

Butadiene. Cis-trans-polymerization[54] of liquid butadiene monomer is
possible in the presence of H-form alumino-silicate obtained by treating a mixture
of mordenite and clinoptiolite with acid, a trialkyl aluminium compound and nickel
acetyl acetone. 8–12 mesh natural zeolites are converted into acid form by treatment with 1 N HCl for 50 hours at 70 °C. A mixture containing about 0.1 g of this
activated zeolite, 0.1 ml aluminium trialkyl in toluene and 2 mg nickel acetyl
acetone, on being heated for 1 hour at 70 °C and finally cooled at −78 °C produces
polybutadiene at room temperature (*cis*-1,4, 94.5 %, *cis*-1,2, 0–99 % and *trans*-1,4,
5.6 %).

1-Propylnaphthalene. Cobalt modified and also nonmodified[55] NaY catalysts can
induce the polymerization of 1-propyl naphthalene. However the mechanism of this
polymerization is obscure and probably involves isomerization reaction.

III Anionic Polymerization of Vinyl and Related Monomers

Molecular sieves have in a limited number of cases been found to exert a definite
influence on the course of anionic polymerizations. However in the few instances
reported these molecular sieves do not participate directly in the initiation reaction.

Lactams are polymerized[56] in solution above their melting point in presence
of a strong base, a molecular sieve and an initiator to low molecular weight material
in low conversion. Thus caprolactam polymerizes at 80 °C in presence of sodium
hydroxide, nalsit 4A and N-acetyl caprolactam to yield nylon 6 (degree of polymerization 480 containing 3 % low molecular weight product). Similar polymerizations have been reported for ω-caprylolactam[56], ω-enantholactam[56],
ω-laurolactam[56] and bicaprolactam[56]. Molecular sieves with effective pore diameter 3–6Å[57] have been used for the continuous removal of water and alcohols in
the polymerization of lactams[57]. An organic clay catalyst[58] prepared by ion
exchange reaction of clay with compound having ≥ 1 amino group polymerizes
caprolactam to a polymer containing 37 % of polyamide in 6 hours at 250 °C.

IV Miscellaneous Applications of Molecular Sieves in Polymerization and Post Polymerization Reactions

A. Vulcanization and Crosslinking

Dehydrated and nondehydrated[59] synthetic zeolites of the types CaX, CaA, MgA, NH$_4$A etc have been used as vulcanization agent for chloroprene rubber[59]. The zeolites abstract hydrochloric acid and form bonds with the metal chlorides resulting in crosslinking in chloroprene rubber. Incorporation of 2 parts zeolite CaX as a vulcanizing agent affords stability towards scorch and vulcanizates with very high wear resistance and stability to the aggressive media. Vulcanization of poly-chloroprene rubber[60] and the same compounded with SKN–18 (copolymer of butadiene with 17–20 % acrylonitrile) by compounding with synthetic zeolites NaA and channel black, saturated with hydrogen sulphide gas for 20 minutes results in materials as strong as rubber vulcanized with sulphur in presence of metal oxides with better resistance properties to petroleum fractions, oils, concentrated hydro-chloric acid, oleic acid, fresh water or sea water, or 70 % sodium hydroxide solution.

High density polyethylene[61] may be crosslinked at the processing temperature by the introduction of powdered zeolite NaX containing dicumyl peroxide or di-tert-butyl peroxide. However, no significant difference has been observed in the strength properties in the crosslinked high density polyethylene and the same containing an equivalent amount of zeolites but elongation at break of crosslinked high density polyethylene (500–600 %) exceeds that of the controlled samples (20–30 %).

Zeolites modified with organic solvents[62] increase the resistance and degree of radiation crosslinking of filled poly-vinyl-chloride. The dehydrochlorination rate of poly-vinyl-chloride on γ-irradiation decreases in the presence of zeolites due to the sorption of hydrochloric acid by the filler. Modification of zeolite with carbon tetrachloride and methylene chloride decreases the swelling degree of filled poly-vinyl-chloride also.

B. Improvement of the Properties of Product Polymer in Presence of Molecular Sieves

Several interesting observations have been made recently which indicate that mole-cular sieves can modify the polymers to yield products with perceptibly improved properties. In this context, it is interesting to note that the molecular weight of polyester and polyamide fibres[63] may be increased and stabilized resulting in polymers with improved mechanical properties. In a typical case[63] about 0.074 g zinc salt of molecular sieve A on being heated for 4 hours at 450 °C cooled, and stirred with 6 g of polyethylene terephthalate for 5 hours at 70 rph and 270 °C under dry nitrogen flow rate 4 1/hour yields polymer with $[\eta] = 0.955$ dl/g and molecular weight 29500. Corresponding values prior to such treatment are 0.61 and 17,000 respectively.

CaX zeolites[64] improve the compactness to the rubber products vulcanized without excess pressure.

Incorporation of 0.1—4% zeolites[65] 13X, 5A, 4A, 10X, XW and also natural zeolites and synthetic polymers (polystyrene, vinyl chloride-vinyl acetate co-polymer, poly propylene, acrylonitrile, butadiene styrene copolymer, poly (methyl methacrylate or polyethylene) containing 0.1—4% antistatic agent improves anti-static properties. However the zeolites alone fail to do so[65]. In another in-stance[66] a composition of polyvinylchloride 100, dioctylthalate 80, stabilizer 2, Pd-stearate-1, and zeolite 100 parts is rolled at 160 ° and pressed to give a white sheet having surface resistivity 3.8×10^8 ohm·cm compared with 1.5×10^{13} ohm·cm for a similar sheet containing calcium carbonate in place of zeolite, which reflects the definite role of these zeolites in improving the antistatic properties of the composition.

C. Inhibition of Polymerization

In a single instance[67], it has been reported that molecular sieves treated with organic compounds containing oxygen in the form of OH-groups or ether linkages inhibit the polymerization initiating power of these zeolites towards unsaturated compounds.

V Concluding Remarks

Molecular sieves find extensive applications as polymerization catalysts mainly for the cationic polymerization of miscellaneous vinyl monomers. In a few cases high molecular weight polymer formation has been possible with these catalysts. In a limited number of instances there is evidence that molecular sieves exert some specific influence on the anionic polymerization of the lactams. However, barring a few cases, mechanisms of all these polymerizations are not unambiguously under-stood, apparently calling for further studies along these lines.

A second potential use of these sieves is as polymer modifiers and crosslinking agents. Improvement of polymer properties after modification in presence of the molecular sieves is possible.

Three important aspects of these sieve-catalysed polymerizations appear to have been neglected.

1) There is no specific study on the regeneration of the sieve catalysts after their use as polymerization catalysts. Consequently, the effect of such regeneration on the catalytic efficiency of these sieves is not known although this information is of much relevance in respect of the commercial application of these sieves as poly-merization catalysts.

2) No information is available either as to the applicability of these sieves as components of stereoregulating catalyst systems.

3) A very significant feature of these catalysts is the modification and improvement of their catalytic properties after appropriate metal loading. Suitable metal loading of these sives may furnish more useful and efficient polymerization catalysts and thus expand the scope of these molecular sieve catalysts.

Information along the line indicated as above is essential for a successful and broader utilization of these sieves as effective polymerization catalysts. Some work in these directions has already been initiated in the author's laboratory but additional work is apparently warranted.

Acknowledgements. The authors express their thanks to Prof. A. Ledwith for his encouragements and to the authorities of the Indian Institute of Technology, Kharagpur for facilities. Financial support in the form of a research scheme by the Council of Scientific and Industrial Research, India is also gratefully acknowledged.

VI References

1. Berl, W. G.: Physical methods in chemical analysis, New York – London: Academic Press, 1961, Vol IV, p.48
2. Panaiotov, Iv., Dimitrov, Iv.: Izv. Inst. org. khim., Bulg. Akad. Nauk. *3*, 35–42 (1967), through Chem. Abstr. *68*, 59951 (1968)
3. Barson, C. A., Knight, J. R., Robb, J. C.: Brit. Polym. J. *4*, 427–435 (1972)
4. Rychly, J., Lazar, M.: J. Polymer. Sci. *B7*, 843 (1969)
5. Ebdon, J. R.: Brit, Polym. J. *3*, 9 (1971)
6. Bertsch, L., Habgood, H. W.: J. Phys. Chem. *67*, 1621 (1963)
7. Hirschler, A. E.: J. Catalysis. *2*, 428 (1963)
8. Benson, S. W., Chaudhuri, A. K., Bittles, J. A.: J. Polym. Sci. *A2*, 3203 (1964)
9. Wolf, F., Bergk, K. H.: Chem. Abstr. *83*, 59440 (1975)
10. Charles, E. S., Wesley, R. C.: Chem. Abstr. *64*, 11398 (1966)
11. Barrer, R. M., Oei, A. T. T.: J. Catalysis. *30*, 460–466 (1973)
12. Barrer, R. M., Oei, A. T. T.: J. Catalysis. *34*, 19–28 (1974)
13. Biswas, M., Maity, N. C., Laha, D.: Unpublished
14. Biswas, M., Chakravorty, D.: J. Polymer Sci., Polymer Chem. Ed., *11*, 7 (1973); Biswas, M., Mishra, P. K.: J. Polymer Sci., Polymer Sci. Polym. Letts. Ed. *11*, 639 (1973); Biswas, M., Kamannarayana, P., J. Polym. Sci. Polym. Chem. Ed. *13*, 2035 (1975); Biswas, M., Kamannarayana, P., J. Polym. Sci. Polym. Chem. Ed. *14*, 2071 (1976)
15. Liengme, B. V., Hall, W. K.: Trans, Faraday Soc. *62*, 3229 (1966)
16. Venuto, P. B., Hamilton, L. A., Landis, P. S.: J. Catalysis. *5*, 484 (1966)
17. Turkvich, J., Nozaki, F., Stamires, D.: Proc. 3rd Int. Cong. Catalysis. *1*, 586 (1964)
18. Venuto, P. B., Landis, P. S.: Adv. Catalysis. *18*, 259 (1968)
19. Nishizawa, T., Hattori, H., Shiba, T., Uematsu, T.: Chem. Abstr. *77*, 88943 (1972)
20. Mortikov, E. S., Minachev, Kh. M., Leontev, A. S., Masloboev-Shvedov, A. A., Kononov, N. F., Lipikhin, M. P.: Chem. Abstr. *79*, 19439 (1973)
21. Lapidus, A. L., Maltsev, V. V., Garanin, V. I., Minachev, Kh. M., Eidus, Ya. T.: Chem. Abstr. *84*, 73561 (1976)
22. Hara, N., Yashima, T.: Chem. Abstr. *84*, 31761 (1976)
23. Yashima, T., Nagata, J., Shimazaki, Y., Hara, N.: Chem. Abstr. *86*, 190505 (1977)
24. Bruk, M. A., Abkin, A. D., Demidorich, V. V., Eroshina, L. V., Urman, Y. G., Slonim, I. Va., Ledeneva, N. V.: Chem. Abstr. *82*, 156766 (1975)
25. Norton, C. J.: Chem. & Ind. (London) 1962, pp. 258–9
26. Norton, C. J.: Ind. Eng. Chem. Process Design and Development. *3*, 230 (1964)
27. Bourne, K. H., Metcalfe, C. J. L.: Chem. Abstr. *69*, 43387 (1968)

28. Shauki, M. Kh., Panchenkov, G. M. Kuznetsov, O. I.: Chem. Abstr. *75*, 77333 (1971)
29. Shauki, M. Kh., Panchenkov, G. M., Kuznetsov, O. I.: Chem. Abstr. *74*, 111488 (1971)
30. Eidus, Y. T., Lapidus, A. L., Rudakova, L. N., Isakov, Y. I.: Chem. Abstr. *80*, 120104 (1974)
31. Lafer, L. I., Nabiev, B. A., Yakerson, V. I.: Chem. Abstr. *84*, 9235 (1976)
32. Driscoll, G. L., Hirschler, A. E.: Chem. Abstr. *76*, 73053 (1972)
33. Bielanski, A., Datka, J., Drelinkiewicz, A., Maleckev, Anna.: Chem. Abstr. *85*, 10631 (1976)
34. Barrer, R. M.: J. Soc. Chem. Ind. (London). *64*, 133–5 (1945)
35. Barrer, R. M., Brook, D. W.: Trans. Faraday. Soc. *49*, 940 (1953)
36. Barrer, R. M., Kravitz, S.: Chem. Abstr. *70*, 46724 (1969)
37. British Petroleum Co. Ltd.: Chem. Abstr. *59*, 11241 (1963)
38. British Petroleum Co. Ltd.: Chem. Abstr. *60*, 1519 (1964)
39. Gensheimer, D. E., Brown, E. C.: U. S. Pat. 3061654 (1962)
40. Rhein, R. A., Clarke, Jeanne S.: Polymer. *14*, 333 (1973)
41. Breck, D. W., Flanigen, E. M.: Molecular Sieves, London: Society of Chemical Industry 1967
42. Barrer, R. M., Villiger, H. Z. Kristallog. *128*, 352 (1969)
43. Linde Molecular Sieves, Tech. Bull. F-1979 B Union Carbide Corp., Linde Div., New York
44. Breck, D. W.: J. Chem. Educ. *48*, 678 (1964)
45. A Report on Molecular Sieve Catalysis, Bull. F–1578, Union Carbide Corp., Linde Div., New York (1967)
46. Bull. F–2808, Union Carbide Corp., Linde Div., New York (1967)
47. Lapidus, A. L., Isakov, Ya. I., Rudakova, L. N., Minachev, Kh. M., Eidus, Ya. T.: Chem. Abstr. *84*, 43196 (1976)
48. Lapidus, A. L., Isakov, Ya. I., Minachev, Kh. M., Eidus, Ya. T.: Chem. Abstr. *85*, 192138 (1976)
49. Paule, J., Anne, M. C.: Chem. Abstr. *56*, 6171 (1962)
50. Goodman, M., Abe, A.: J. Polym. Sci. *A2*, 3471 (1964)
51. General Electric Co.: Chem. Abstr. *67*, 10311 (1967)
52. Amemiya, T., Kurokawa, K.: Chem. Abstr. *68*, 3347 (1968)
53. Bamford, C. H., Eastmond, G. C., Maltman, W. R.: Trans. Faraday Soc. *61*, 267 (1965)
54. Kawasaki, A., Taniguchi, M., Nishiyama, T.: Chem. Abstr. *75*, 37621 (1971)
55. Dimitrova, R., Dimitrov, Kh.: Chem. Abstr. *84*, 17010 (1976)
56. Zavody, A., Zapatockeho, N.: Chem. Abstr. *75*, 36995 (1971)
57. Kralicek, J., Kubanek, V., Solcova, J., Kondelikova, J.: Chem. Abstr. *77*, 6031 (1972)
58. Fujiwara, S., Sakamoto, T.: Chem. Abstr. *86*, 141002 (1977)
59. Rapchinskaya, S. E., Blokh, G. A., Khromenko, S. P.: Kauch, Rezina, *27* (9), 17–20 (1968)
60. Nosnikov, A. F., Blokh, G. A., Esman, P. I., Petrova, T. K., Chernisskaya, T. V., Lazurovich, T. Ya.: Kauch, Rezina. *27* (3), 18–21 (1968), through Chem. Abstr. *69*, 11220 (1968)
61. Proskurnina, N. G., Akutin, M. S., Budnitskü, Y. M.: Chem. Abstr. *82*, 140928 (1975)
62. Dakin, V. I., Nikolaev, V. I., Egorova, Z. S., Karpov, V. L., Chebanyuk, S. A.: Chem. Abstr. *84*, 122710 (1976)
63. Wolf, F., Rollin, J.: Chem. Abstr. *80*, 60956 (1974)
64. Khvastunov, A. A., Popov, A. V., Blokh, G. A.: Chem. Abstr. *83*, 133012 (1975)
65. Tokyo Shibaura Electric Co. Ltd.: Chem. Abstr. *68*, 22458 (1968)
66. Nishikawa, Y., Canon, K. K.: Chem. Abstr. *84*, 165737 (1976)
67. Kamn, G. R.: Chem. Abstr. *78*, 18634 (1973)

Received July 27, 1978
A. Ledwith (editor)

Modified Polyethylene Terephthalate Fibers

Luigi Szegö

The performance of standard poly-ethylene terephthalate filaments is in many cases seriously prejudiced by some negative properties of these fibers, such as their nearly complete water repellency, difficult dyeability, tendency to pilling, formation of static, etc. The present paper aimes to recall the methods proposed and realized in order to overcome these shortcomings by modification of the macromolecule. The subject is divided in two groups regarding respectively the modifications for obtaining changes of mechanical properties and shape of the fiber (shrinkage, elasticity, pilling performance, impact resistance, crimping) and of specific properties (dyeing behavior, static load, flame resistance, soiling and soil removal).

Table of Contents

1 Introduction

Polyethylene-terephthalate (PET) fibers, prepared according to traditional methods, are among the most stable synthetic fibers, owing to their outstanding physical and chemical resistance and their complete indifference to moths, mildews, and microorganisms. These characteristics are due to the high degree of orientation and structural order of macromolecular chains.

These excellent characteristics are, however, also the source of some negative features such as nearly complete water repellency, difficult dyeability, tendency to pill, and formation of static, etc. Many of these shortcomings have slowed down or even made impossible the use of PET fibers in certain textile areas and have made the fiber producers aware of the necessity of meeting more exacting consumer demands. The upshot of these circumstances has been the manufacture of so-called modified fibers[1].

Modification is generally considered to be a change in the chemical composition of the PET molecule, achieved by introducing into the chain small quantities of a third and perhaps even of a forth component.

In this connection it should be borne in mind that structural modification of the fiber is a double-edged weapon since the improvements are often achieved at the expense of existing fiber properties. The newly modified textile material may definitely show some valuable new properties for the required performance, but at the same time some of its excellent original features may have become impaired so as to exclude the modified fiber from some of the traditional end uses of the PET material. Modification of the macromolecule should thus be achieved with moderation in order to ensure real success in improving fiber quality at least in a given textile area. This fact, explains why the modifications have resulted in different effects for different ends and why the improvements concern only one or a few properties of the fiber. For this reason, the big producer companies put on the market a great number of modified PET fibers, each of which is only capable of satisfying the requirements for specific performances.

2 Modifications for Obtaining Changes in Mechanical Properties and Fiber Shape

2.1 High-Shrinkage Fibers

A very interesting PET fiber endowed with a high-shrinkage property is now manufactured on an industrial scale (e.g. Hoechst's "Trevira 550").

This fiber, for which a significant growth rate can be foreseen, may be used in the textile area in blend with normal PET fibers, wool, or cotton for preparing bulky, soft yarns for men's and ladies' wear and for felts. This kind of fiber must possess high shrinkage when heated dry or in medium-hot water as well as high tenacity, but not too high elongation at break. For textile uses these parameters are included in the following limits: denier = 1.0–4.5, tenacity about 4 g/d, elongation at break = lower than 60%, hot water shrinkage = 30% – 50%. For use in technical areas, fibers are recommended with denier = 1.0–1.5 and boiling-water shrinkage, 40%–60%.

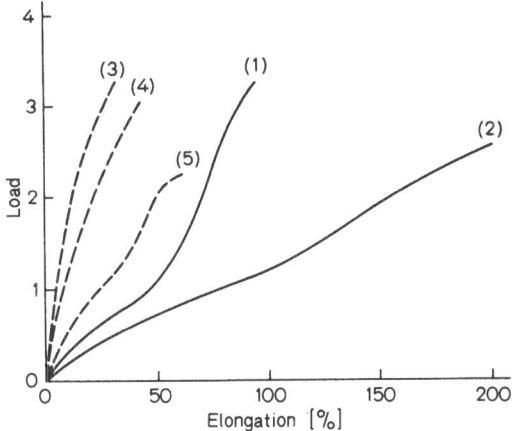

Fig. 1. Load/extension curves of highly shrinkable PET fibers: (1) PET homopolymer not treated; (2) PET homopolymer hot-water treated; (3) PET copolymer for textile uses, not treated; (4) PET copolymer for textile uses, hot-water treated; (5) PET copolymer for textile after treatment, with hot air (200 °C)

Under thermal treatment, this fiber shrinks, causing the other fibers of the blend to crimp. In this way the shrunken fibers become the bearing component in the yarn. Since they form 30%–40% of the blend, their tenacity should remain elevated after shrinkage. Besides the amount, the force of shrinkage also must be considered, the high-shrink fiber having to overcome a considerable mechanical resistance in the yarn and/or in the fabric[2−3].

It is known that residual shrinkage is strictly related to the degree of crystallinity of the fiber, high shrinkage being linked to low crystallinity. It should, however be borne in mind that shrinkage gives rise to new crystallite formation, since the increase of denier without a proportional growth in diameter of the filaments means higher density, i.e., a tighter bundling of the macromolecular chains.

The simplest and more economical method should be to use homopolymer PET for the manufacture of high-shrinkage fibers. Technically, nothing prevents the use of homopolymer PET filaments of a high degree of orientation and a low degree of crystallinity, produced for example at high spinning speed, high deformation ratio,

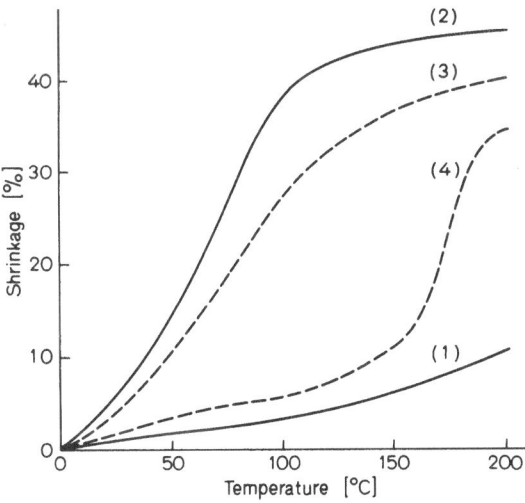

Fig. 2. Shrinkage of different PET fibers in boiling water and then in hot air (up to 200 °C: (1) PET homopolymer, standard type; (2) PET homopolymer, high-shrinkage type; (3) PET copolymer, for knitted goods; (4) PET copolymer, for woven goods

low denier, and eventually drawing at low temperature and free-shrinking the fibers with hot water by 40% or more. The fact that under these conditions shrinkage occurs at 100 °C and that shrunken filaments have very high elongation at break (see Figs. 1 and 2) makes them unsuitable for textile uses.

To be convenient for the textile area, a high-shrinkage fiber must show a tensile behavior like that of normal PET fibers and almost no shrinkage up to 100 °C, but considerable shrinkage at higher temperatures so as to be dyeable when boiled in the loose state, sliver, or yarn — eventually on cops — without shrinkage, the total shrinkage being developed during heat-setting of finished fabrics at about 190 °C. To obtain this result fibers should be prepared under conditions so as to be completely amorphous or slightly crystalline and highly oriented. This cannot be achieved by using homopolymer PET, but only by using *copolymers* in which a certain percentage of the normal components, e.g., terephthalic acid and ethylene glycol, is substituted by an other bicarboxylic acid and/or diol.

There are many papers on the subject[4−8], which shall be briefly summarized. In order to influence positively the amount and the force of shrinkage through a chemical modification of the macromolecule, many compounds have been proposed, a selection of which is listed in Table 1.

The co-monomers statistically embodied in PET cause an increase in the crystallization temperature and consequently a minor crystallization rate, temperatures being equal. There is a close relationship between the modifying effect of these compounds and their concentration in the polyester, while the force of shrinkage depends heavily on the nature of the co-monomer itself. According to some authors[9−10] the co-monomer selected should induce the formation of an angle in the chain. From this point of view isophthalic acid has a marked modifying effect, the maximum being at about 10% by weight of the PET.

In manufacturing the copolyester, it should be borne in mind that the polycondensation rate is noticeably lower than that of homopolymer PET and provisions should be made for this effect. The melting point of the modified fiber is about 20 °C lower and the resistance to chemicals somewhat worse than that of standard PET fibers.

In the technical area, the high-shrinkage fibers are used primarily in the preparation of felts, which are used as supporting layers in the manufacture of synthetic leather.

Table 1.

Bicarboxylic acids	Diols
Isophthalic acid	Higher or substituted glycols
5-Methyl-isophthalic acid (uvitinic acid)	Example:
5-Oxy-isophthalic acid	2,2-dimethyl-1,3-propandiol
2,5-Dichloro-terephthalic acid	(Neopentyl glycol)
4,4-Diphenyl-dicarboxylic acid	
2,5-Furan-dicarboxylic acid	
Adipic acid	
Sebacic acid	

In the textile area, the high-shrinkage fibers yield excellent results in blends for producing bulky yarns and fabrics for overcoats and men's and ladie's wear. The feel and drape of the fabrics may be considerably varied by using different fibers in the blend (wool, polyacrylics, polyamides). The process can be greatly improved by arranging the high-shrinkage fibers in the blend, e.g., moving them to the middle of the yarn in order to make the most of their elevated tensile properties.

Another important advantage of high-shrinkage fibers lies in the fact that the bulk of the goods does not change, whereas that of texturized fibers and yarns decreases after a certain amount of use and wear.

Spontaneously extensible filaments, which may be considered to be the opposite of shrink fibers are useful, e.g., in sewing thread in order to prevent seam puckering, but may also be employed to increase the bulk of yarns by plying these filaments with standard or shrink fibers. To make fibers that have the property of spontaneous extensibility when heated in hot water, Du Pont discloses in a patent[11] some working methods, which can be summarized as follows.

Oriented filaments of PET (intrinsic viscosity (η) = 0.5–0.65) spun at high speed and substantially amorphous, are stretched slightly (draw ratio less than 1:3) so as to maintain their crystallinity degree as low as possible (less than 5%). Such filaments are treated in the relaxed state with water at 70–75 °C for a short time (about 5 min). In this way a shrinkage of 30%–50% takes place and the orientation degree diminishes considerably while crystallinity increases to 20%–30%. These filaments are in metastable equilibrium. In fact when treated once more with hot water (90–100 °C) they increase in length irreversibly by 1%–2% or more. Copoly-

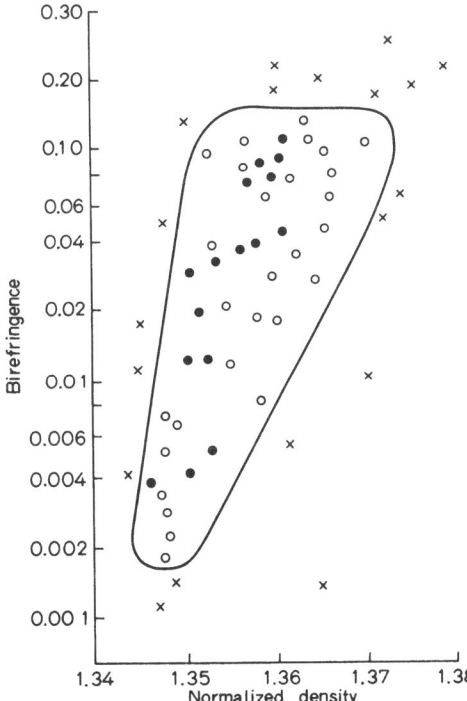

Fig. 3. Birefringence vs. normalized density of polyester fibers[11]. (a) spontaneously extensible by hot water treatment: ○ homopolymer PET fibers; ● copolymer PET fibers; (b) x shrinkable in hot water

esters mentioned in this connection are formed by substituting terephthalic acid in
PET by 15 mol % sebacic, isophthalic, or hexahydroterephthalic acid.

This unusual behavior depends on a novel combination of molecular orientation (as measured by birefringence) and crystallinity (density) achieved by the first
thermal treatment. From several tests it was seen that only filaments having birefringence values in the range of 0.0015–0.15 and a "normalized" density within
the area enclosed in the graph shown in Fig. 3, are spontaneously extensible when
treated with hot water. Normalized density is calculated by the formula
$\rho_N = \rho + (1.3400 - \rho_A)$ where ρ is the observed density of the fiber and ρ_A is the
observed density of a completely amorphous specimen.

2.2 Elastomeric Fibers

Today, most elastomeric fibers are formed by polyurethanes. These fibers are expensive owing to the great number of stages needed for their synthesis and to the
necessity of spinning the solution of the polymer rather than the melt.

For this reason many attempts have been made to modify polyamides and particularly polyesters in order to improve their elastic behavior. It is necessary, however, to recall here that this problem is far from being resolved, so that the following
information is given only in order to complete the list of PET fibers having modified
mechanical properties.

The first report on PET fibers of increased elasticity, achieved by partial substitution of ethylene glycol by polyethylene glycol (PEG) or by tetrahydrofuran,
goes back to 1958[12–14]. It was reported then that a copolymer containing 40%–60%
polyethylene glycol (mol. wt. 4000) has an elongation at break of about 350%.

The preparation, spinning, and drawing of copolymers PET–PEG were investigated more recently[15–16] and it was ascertained that by incorporation under optimum conditions of about 5 mol % PEG (mol. wt. 4000) into PET, the polyglycol
reacts with the terephthalic group and the copolymer obtained can be easily spun
and drawn. However, fibers prepared in this way are far from having the elasticity
of polyurethane fibers (e.g. Lycra) (Table 2).

Slightly better results are reported in a patent of Asahi Chem. Ind.[17], in which
a polycondensate, formed by PET and a copolymer from 17% ethylene oxide and
83% propylene oxide of molecular weight 1000, is described. The copolyester
(mp = 160 °C) is spun at 190 °C and, according to the example reported in the
patent, filaments of 100 Den (diam = 0.3 mm) obtained in this way have a tenacity

Table 2.

Fiber	Tenacity g/d	Elongation at break (%)	Elastic recovery (%) after elongation of				
			50%	100%	200%	300%	400%
Lycra copol.	0.6–0.9	500–600	100	97	95	94	90
PET/PEG	2.0–2.5	150–200	80	–	–	–	–

of 0.45 g/d, elongation at break of about 350%, and elastic recovery of 85% at 100% elongation.

Actually the probability of producing elastic fibers from PET seems rather low, since attempts thus far have yielded poor results.

A further point to be considered is the particularly low heat stability of these copolymers, due to the frequent ether bonds in the chain.

2.3 Low-Pilling Fibers

The appearance of small pills of entangled fibers on the surface of soft fabrics and knit goods after a certain amount of wear, commonly called "pilling", has a considerable negative effect on the quality of the garment. This defect became apparent on a large scale with the extensive use of synthetic fibers in the textile industry. The "pills" are built up from fibers from the surface of the textiles, which come out and become entangled. While the entanglements formed form natural fibers disappear readily due to wear abrasion, pills formed from synthetic fibers tend to remain stuck to the textile because synthetic fibers possess a very high flexing resistance, similar to that of polyesters. The mechanism of pilling, as well as both chemical, physical and textile provisions for avoiding this defect, are discussed at length in a series of papers[18-23].

For improving the resistance to pilling, two methods can be used based on the following concepts:

A. Inhibition of the Coming out of Fibers:

The coming out of the fibers from woven or knit goods can be slowed down by changes in the textile variables, e.g., by
a) an increase of the fiber *length,*
b) an increase of the *denier;* this variation is, however, limited by the hand and the drape of the good, highly sensitive to fiber diameter,
c) a modification of the fiber *surface,* which has a marked influence on pilling, fibers with noncircular cross section having an improved performance,
d) an increase of *twist,* which is very effective for inhibiting the pilling,
e) an increase of insertions and general modification of the fabric weave can improve the pill resistance, though this measure is not applicable for loose and soft textiles.

In the latter cases the only possible way of preventing pilling is to use fibers with diminished resistance to flexing.

B. Promoting the Removal of Pills by Lowering the Flexing Resistance of the Synthetic Fiber:

Two different concepts can be applied to achieve this aim:
a) decrease of the molecular weight of homopolymer PET,

b) incorporation into the chain of monomers, different from PET repeating units,
   with building up of copolymers.

1) Staudinger[24] was the first to prove the link between degree of polymerization and mechanical properties, showing for example that after hydrolytic degradation of cotton from $\bar{P}$ 1000 to 500, tenacity drops by 30%, elongation at break by
32%, while flexing resistance decreases by 91%. According to Thimm[25], the same
thing happens to PET fibers of equal denier and elongation at break by varying the
degree of polymerization, as illustrated by the data relating to PET fibers of the
Hoechst Co. listed in Table 3.

By sacrificing a relatively small percentage of tenacity (e.g. 30%), flexing resistance can be diminished by 90% and pilling by about the same value. The fact that
approximately the same relationship exists between these data and those found by
Staudinger on cotton, sheds significant light on this phenomenon.

For the producer of fibers who uses this process, the trouble begins with a strong
drop of the melt viscosity η_0 due to the lower molecular weight and the intrinsic
viscosity of the polymer. Since, for regular spinning of PET, it is a requirement that
melt viscosity should not be less than 1000 at 280 °C, a process other than that based
on the decrease of molecular weight may be more convenient.

2) The process based on the incorporation in the chain of a certain number of
monomers, different from the PET repeating units, produces a chemical modification by interrupting the sameness of the macromolecule, the main changes being[26]:
a) an *angle* is formed in some point of the chain, using for example isophthalic instead of terephthalic acid and thus causing a flexing of 60 °C of the more or less
   straight polymer,
b) monomers with *short lateral chains* are embodied in the polyester. For this purpose, 2,2-dimethyl-1,3-propandiol (Neopentylglycol) is frequently used as a partial substitute for ethylene glycol.
c) *Large bi- and tridimensional branches*[27–30] are grown by using polyvalent alcohols such as glycerol, trimethylolpropane, pentaerythrite, etc., in partical substitution of ethylene glycol and/or polycarboxylic acids such as trimelytic acid,
   in partial replacement of terephthalic acid, modifying in this way the morphologic
   order of the polymer.

Using these methods, the melt viscosity undergoes significant changes, when
molecular weight and temperature are held constant. The same is also true for the
mechanical properties of the fiber especially the flexing resistance. The various pro-

Table 3.

Type	Mol. wt.	$[\eta]$ dl/g	η_0 (280 °C) Poise	Ten. g/dtex	Flex. resist. (revolutions)	Pilling rating (Reutling scale)
220	10,000	0.63	4800	4.3	3000	–
330	14,000	0.51	900	3.8	2000	6.6
360	11,400	0.44	700	3.3	900	4.3
340	11,200	0.42	600	3.2	800	–
350	10,500	0.40	390	3.0	300	0.8

cesses can also be combined. Actually many patents claim both a decrease of molecular weight and a structural change in order to improve the pilling performance.

Low-pilling fibers are commonly used on a large scale especially in blends with cotton, wool, and other synthetic fibers, for the manufacture of soft fabrics, knits, pile, and in other areas.

2.4 Impact-Resistant Fibers

It is well known that by increasing the molecular weight of PET, yarns with improved properties can be produced that are suitable for tire cords, their tenacity not being inferior to that of polyamide yarns used in this area. At the same time, attempts have also been made to improve the impact resistance of PET yarns, this factor being of considerable importance in the construction of airplane tires.

From the scanty reports published on the subject, it seems that by using certain terephthalic copolymers, especially by embodying suitable elastomers in the chain, improvements in impact resistance may be achieved.

For example[31], it is claimed that a copolymer can be produced from dimethyl terephthalate, 1,4-cyclohexanedimethanol and dibutyl-p,p'-sulfonyldibenzoate by following the usual polycondensation procedure. Yarns manufactured from this copolymer are said to have an improved impact resistance.

In two patents[32-33], the incorporation into PET of 10% Hycar 1042 is described, i.e., an elastomer formed by equal parts of nitrile rubber and polyethylene or of 5% Hostalon APK, which is a rubber made from an ethylene-propylene copolymer. It is claimed that in this way impact resistance is increased by 25%–50%.

2.5 Texturized Fibers

All of the PET fibers and a majority of the continuous filament yarns are texturized, i.e., crimp is applied.

This operation can be achieved by different methods:
A) *Mechanical crimping* is carried out by false twisting, compression in a stuffing box, knitting, edge bending, by gears, and hot blows[34-35].
B) *Dissymmetric structure* of fibers is obtained by spinning two component filaments side by side or utilizing the different swellings of the layers formed by the core and skin present in polyester filaments.
C) *Dissymmetric heat treatment* is carried out by making filaments slide on hot plates at a temperature somewhat higher than the melting point of PET or placeing the filaments in contact with cold plates immediately under the spinneret.

Processes reported in (A) are common to all synthetic fibers and shall not be discussed here. According to recent papers, processes (B) and (C) are more specific for PET filaments. Without pretending to be complete, some of the patents issued on the subject are summarized here.

B.1. The crimp of two-component filaments grows according to the different degrees of shrinkage of the two parts forming the cross section. In this connection, the following combinations have been proposed:

Homopolymer PET/copolymer formed by PET and pentaerithrolterephthalate
Homopolymer PET/copolymer formed by PET and 10 mol% polyethylene-
isophthalate. The yarn is stretched at temperatures between 70 and 170 °C
and relaxed at 180 °C[36–37].
Homopolymer PET/copolymer formed by 85% PET and 15% polyethylene-
hydroxy-benzoate. Filaments are stretched by 430% at 90 °C and crimp is developed at 150 °C[38].
Homopolymer PET/5%–20% Polystyrene[39].

All these processes are difficult to accomplish in practice, owing to the sophisti-
cated equipment required to lead the two molten jets to the same spinneret hole. For
this reason processes based on another concept, especially on the different swellings
of layers of various structures, have been proposed. By developing the core/skin
fiber structure and making it asymmetric, this aim has been more or less achieved.
The following examples illustrate this procedure.

B.2. PET filaments are treated for 9 min at room temperature with a 10:1
solution of hexafluoroacetone: water. Crimp is developed owing to the different
degrees of swelling of the core and skin layers[40].

Nearly amorphous filaments of PET ($(\eta) = 0.57$), spun at 295 °C at 100 m/min,
are dipped in boiling water, and stretched 1:2.5 at a rate of 500 m/min. By treating the filaments in the relaxed state with water at 70 °C, crimp is formed[41].
Undrawn filaments of PET are swollen in water at 30–40 °C and stretched by
600% at about 50 °C. After shrinkage in damp medium, crimp is developed by
steam at temperatures between 130 and 210 °C[42].

Bulk achieved in this way is rather low. Better results are obtained with processes
based on asymmetric thermal treatment of filaments.

(C) In a patent of the Hoechst Co.[43] a device is described that consists of a metallic
cylinder with conical ends, fastened at a certain distance from the spinneret, so
as to compel the filaments to touch the surface for a minimum length, defined
by the empirical formula

$$L_{min} = \frac{V \sqrt{T}}{500} \ [cm]$$

where
T = Denier of drawn filaments
V = Spinning rate (m/min)
By fixing the distance between the spinneret and the contact point of the filaments on the cylinder to 13–20 cm and the temperature of the cylinder to
about 20 °C – the spinning rate being 1400–1800 m/min – filaments are obtained that have different orientation degrees in a radial direction, as can be
seen by measuring the birefringence of the cross section. After stretching for
1:3.0–3.5 and shrinkage at 160 °C, the filaments show 20%–28% crimp, though
with relatively low frequency (3–5 per cm). Using a copolymer in which part
of the ethylene glycol is substituted by neopentylglycol, filaments are produced
with crimp frequency up to 20 per cm.
A similar process[44–45] has been proposed, asymmetric quenching being
achieved by blowing cold air crosswise in the spin duct. When spinning a co-

polymer ethylene-terephthalate-isophthalate (10%), fibers are produced, which after cold and hot drawing and shrinkage at 150 °C, have a crimp frequency of about 7 per cm and a good elastic recovery.

Dissymmetric orientation can also be performed by oblique spinning (20–70°) with respect to the spinneret, drawing by 600% in water at about 50 °C, and heat-setting in a relaxed state at 180 °C. A crimp frequency of more than 20 per cm has been reported[46].

Such procedures may be useful for crimping continuous filament yarns instead of employing false twist, which requires expensive equipment.

Crimping of fibers is generally performed by mechanical processes, i.e., impressing the two after stretching on hot gears or better still by compression in stuffer boxes. However, it is useful to establish an asymmetric arrangement of the fibers during spinning, e.g., by blowing cold air sideways at a short distance under the spinneret. By combining the two effects (asymmetric structure and mechanical texturizing), fibers with high crimp and resiliency are produced[47].

A promising method for obtaining bulky yarns is to blend normal and high-shrinkage fibers. The advantages of this process are described in Section 2.1.

3 Modifications for Improving Specific Properties of Fibers

3.1 Fibers with Improved Dyeing Behavior

One of the first difficulties encountered by producers and customers relates to the very low dyeability of PET fibers. Owing to the chemical and physical structure of the polymer, the dyeing performance of standard polyester fibers creates a number of problems. PET fibers lack the typical dye receptor groups present in other fibers and therefore the only class of dyes employed for polyesters, as for cellulose-acetate rayons, are those called "disperse dyes". However, the depletion of the aqueous solutions of these colors due to the passage of the dispersed particles from the solution into the fiber is very slow, even at boil.

Some progress was made by developing new techniques[48–50], based on the use of "carriers" or dyeing at high temperature (HT and Thermosol process). However, carrier dyeing is expensive and creates new problems, both technical and ecologic, i.e., related to the pollution of effluents limited by the laws of many countries. Moreover, traces of carriers left in the fiber lower the light fastness of the color so that a very careful last wash has to be made. For high-temperature dyeing, special pressure-resistant and rather expensive equipment is needed. Thus many attempts have been made to avoid both the use of carriers and of high dyeing temperatures.

3.1.1 Improved Disperse Dye Level

Dyeing with disperse colors depends on the ability of the dyestuff to diffuse into the fiber and upon the capacity of the fiber to absorb dye. These properties depend on the spinning, drawing, and heat-setting conditions, which, as is well known, are responsible for the internal structure of the fiber. Quantitative effects of these conditions on dyeing are shown in Figure 4 and 5[51, 52].

Dyestuffs are absorbed primarily by the amorphous regions and require an elevated mobility of the molecular chains in these regions. The thermal coefficient of this mobility, responsible for the diffusion rate of the dyestuffs, depends largely on the glass-transition temperature, which increases with rising crystallinity and degree of orientation of the fiber. In fact, drawing and heat setting cause a significant reduction of the rate of dye absorption, which, however, can be increased by introducing certain co-monomers in the PET molecule.

This procedure should be applied with great care in order to prevent excessive lowering of the glass-transition temperature and of the melting point, which causes greater losses of mechanical and technologic properties of the fibers. Modification of the PET molecule therefore should not exceed about 15 mol % of the co-component.

Fibers from modified polyesters with sufficiently high melting points, showing a good dyeability with disperse colors, are manufactured from block-copolymers made from PET and polyalkylene glycols (PAG), i.e., polyethylene – or polypropylene glycols having sufficiently high molecular weight (1000–3000). In fact, the decrease of the melting point depends on the molecular ratio between the ethylene glycol and polyalkylene glycols employed, so that the introduction into the chain of the same amount of high-molecular-weight polyalkylene glycol causes a decrease in the melting point that is slightly less than that due to polyalkylene glycols of lower molecular weight. In contrast, the decrease of the glass-transition temperature depends on the amount and not on the molecular ratio of polyalkylene glycols employed[53].

Fibers from copolyesters containing PAG dye to deep shades in a boiling bath of disperse colors also without carriers. The preparation of such copolyesters is, however, troublesome since at the thermal conditions of polycondensation, e.g., at more than 270 °C, PAG tends to depolymerize so that the block-copolymers become copolymers with statistical distribution of the components[54–55]. In this way the decrease of the melting point of the resulting copolymer cannot be avoided. High

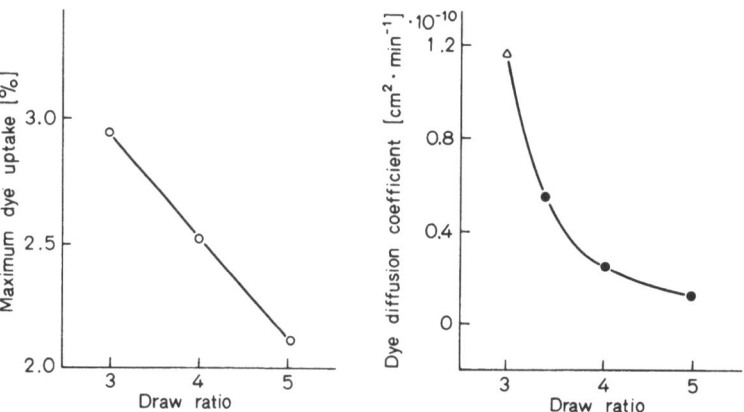

Fig. 4. Dyeing of PET yarns with high affinity disperse dyestuffs vs. draw ratio[51]

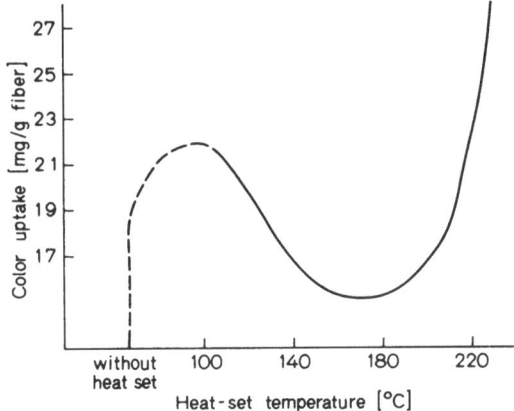

Fig. 5. Disperse color uptake vs. heat-set temperature[52]

vacuum and thin layers of the reacting melt thus become indispensable during poly-condensation.

To prevent or to delay the depolymerization of PAG various measures have been suggested. For example the Teijin Co.[56-59] in a great number of patents, claims that the addition of small amounts of benzene derivates having the general formula:

where R is H or Bu or lauryl propionate or other substitutes, appear to have more or less the same effect.

However, fibers made from these copolymers, have the drawback of being very sensitive to thermal, hydrolytic, and photochemical influences, which is a logical consequence of an amass of ether links present in PAG.

Since results obtained in this way are not satisfactory, incorporation into the polyester chain of other bifunctional compounds was tried with the objective of im-proving dyeability with disperse dyestuffs. Table 4 are contains a list of some of the monomers suggested in patents issued in the last few years.

This list contains only a few of the monomers mentioned in the papers dealing with improvements for dyeing PET fibers with disperse colors. Results, restricted to the monomers listed in the table, which are also convenient from an economic point of view and thus applicable on an industrial scale, are given in Figure 6 and 7. These graphs show the relationship between the melting point or the glass-transition tem-perature T_g, as a function of the amount (mol%) of the monomer incorporated into the polyester. It can be seen that for equal molecular ratios of these compounds in-corporated into the chain of PET, the melting point of the copolyester changes only slightly. Much more remarkable are the differences in the glass-transition temperature due to the type of monomer employed. The graph and extensive practical experience

Table 4.

Dicarboxylic acids	Hydroxy-acids	Diols
$HOOC-(CH_2)_n-COOH$[60-64] n = 3–12	$HO-(CH_2)_5-COOH$[67-68]	$HO-(CH_2)_n-OH$ n = 3–6
$HOOC- \square -COOH$[65]	$HO-CH_2-\overset{\overset{\displaystyle CH_3}{\displaystyle \vert}}{\underset{\underset{\displaystyle CH_3}{\displaystyle \vert}}{C}}-COOH$[71]	$HO-CH_2- \square .-CH_2-OH$[74]
Isophthalic acid[66] $HOOC \cdot CH_2-\phi-CH_2 \cdot COOH$ $HOOC-\phi-CO-\phi-COOH$ $HOOC-\phi-OCH_2-\phi-CH_2-O-\phi-COOH$[69]	$HO-\phi-COOH$[76] $HO-(CH_2)_n-\phi-COOH$ n = 2–3	$HO-CH_2-\overset{\overset{\displaystyle CH_3}{\displaystyle \vert}}{\underset{\underset{\displaystyle CH_3\ CH_3}{\displaystyle \vert}}{C}}-CH_2-OH$ $HO-\phi-\overset{\overset{\displaystyle CH_3}{\displaystyle \vert}}{\underset{\underset{\displaystyle CH_3}{\displaystyle \vert}}{C}}-\phi-OH$[66] (Bisphenol)
$HOOC-CH_2NH \cdot CO-\phi-CO-NH-CH_2-COOH$[70] $HOOC-\phi-OCH_2-CH_2-O-CH_2-CH_2-O-\phi-COOH$[71]		Polyethylene-glycols (m.w. = 400–6000)

Tricarboxylic acids	Polyols
Trimesinic acid[72]	Pentaerythrol[75]

The symbol ϕ is used for brevity to represent the p-phenylene group.

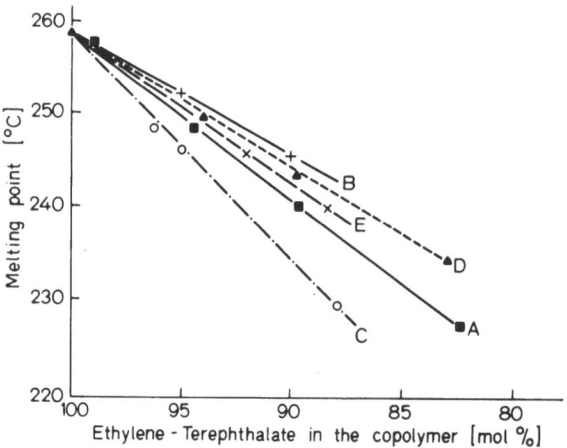

Fig. 6. Effect of the co-monomer on the melting point of the copolyester[76] A : PET/polyethylene-adipate; B : PET/polyethylene-glutarate; C : PET/polyethylene-sebacate; D : PET/polyethylene-isophthalate; E : PET/polycaprolactone

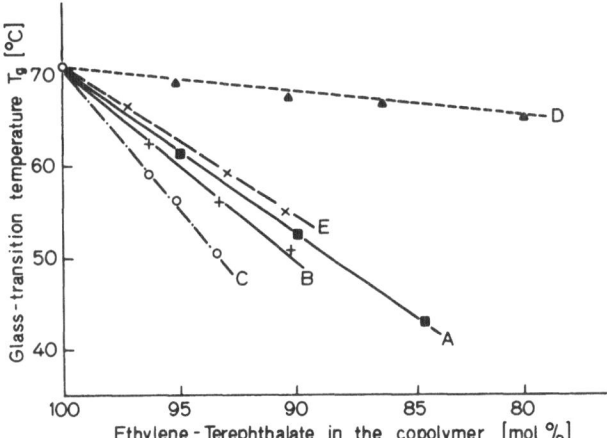

Fig. 7. Effect of the co-monomer on the glass-transition temperature of the copolyester[76]
A: PET/polyethylene-adipate; B: PET/polyethylene-glutarate;
C: PET/polyethylene-sebacate; D: PET/polyethylene-isophthalate;
E: PET/polycaprolactone

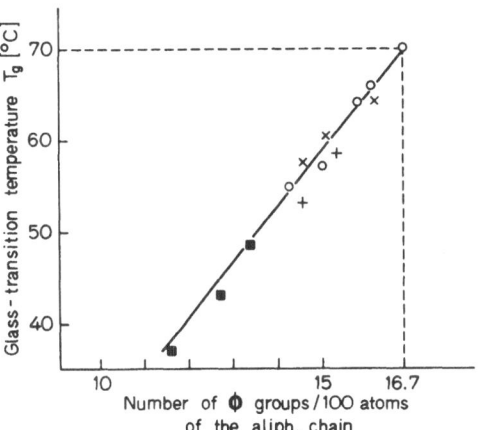

Fig. 8. Glass-transition temperature vs. number of p-phenylene groups/ 100 atoms of the aliphatic chain[77,78]
A o = PET/polyethylene-adipate,
B + = PET/polyethylene-glutarate,
C ▪ = PET/polyethylene-sebacate,
E x = PET/polyeprolactone

indicates that aliphatic, and not branched, compounds cause a greater decrease of T_g than do aromatic components. For instance, a much greater amount of isophthalic acid, in comparison to aliphatic bicarboxylic acids, has to be added to shift the glass-transition temperature by the same value[76].

Polyesters containing only aliphatic, not branched acids, without rings, form very flexible chains and have low glass-transition temperatures. By introducing p-phenylene groups in these polyesters, the value of T_g — according to Edgar[77, 78] — increases as a linear function of the number of p-phenylene groups incorporated for every 100 atoms of the aliphatic chain, attaining the level of $T_g = 70\,°C$ for homopolymer PET, i.e., for the value of 16.67, which corresponds to 1 ϕ for 6 atoms of the aliphatic chain. The higher efficiency of sebacic acid (C_{10}) with respect to adipic acid (C_6) for equal mol % of incorporated monomer is evident from Fig. 8.

The decrease of T_g due to the substitution of terephthalic acid by aliphatic bicarboxylic acids is greater than that achieved by replacing ethylene glycol by other glycols, e.g., by 1,4-butandiol. At any rate, to prevent the improved dyeability from being detrimental to the textile properties of the fiber, the amount of dicarboxylic aliphatic acid used as a modifying agent should be restricted to 5–8 mol % and that of isophthalic acid to 10–15 mol % of terephthalic acid.

Investigations[79] on the synthesis of copolyesters, e.g., on that of polyethylene-terephthalate-cosebacate, indicate that the kinetics of the co-polycondensation is of the same order and that the reaction rate is about the same as that of standard PET. The same holds true for the crystallization rate of copolyesters[80], provided that the amount of co-monomer is limited to the values formerly used, so that the conditions of drawing, heat-setting, and shrinkage are more or less the same as those used in processing PET.

In this way, the tensile properties of yarns and fibers made from copolyesters containing aliphatic bicarboxylic acids remain nearly unchanged in comparison to those of homopolyesters manufactured under the same conditions[81]. Some data are reported in Table 5.

The dyeing behavior of texturized standard and modified PET fibers was tested by two different methods using disperse dyestuffs of high and low affinity toward PET, and the results are given in the same paper[81]. After 1 h of dyeing under the following conditions:

at about 100 °C without and with carrier (o-phenylphenol) of increasing concentration,

at 100 °C and higher temperatures and pressures (HT process), the color depth of the samples was evaluated. From the results shown in Figure 9 it can be seen that the higher dyeing rate of copolyesters makes it possible:

to use baths without or with carriers of reduced concentration, or to dye at 100 °C instead of at higher temperatures and pressures. Only by dyeing at 130 °C is nearly the same color shade observed on standard PET fibers as that observed on copolyesters dyed at boil and atmospheric pressure.

Table 5.

	Yarns		Fibers	
	Homopolyesters	Copolyesters	Homopolyesters	Copolyesters
Denier/filam.	167/32	167/32	1.7	1.7
Tenacity g/d	4.4	4.2	3.5	3.5
Elong. at break %	22	23	67	64
Shrink at boil %	8	12	0	0
Shrink at 200 °C %	15	20	3–4	7–10
After texturizing				
Tenacity g/d	3.3	3.2		
Elong. at break %	27	28		
Crimp %	About 12	about 11		

Comparison of the dyeing behavior of standard polyesters with that of copolyesters confirms that block-copolyesters containing polyalkylene glycols and PET
modified by aliphatic or aromatic compounds show a markedly better dye uptake,
uniformity, and light fastness than standard polyesters, especially the copolyesters
of the latter type.

3.1.2 Dyeable Fibers with Cationic Dyestuffs

From the theoretical constitution of the PET molecule
$$HO \cdot CH_2-CH_2-O-(CO-\phi-COO-CH_2-CH_2-O)_n-CO-\phi-COO-CH_2-CH_2-OH$$
the neutral character of the polyester can be inferred, since only hydroxyl end
groups are present.

However, analysis of end groups shows that even standard PET, prepared from
dimethyl terephthalate or terephthalic acid and an excess of ethylene glycol does
not have this formula. Despite careful processing, secondary reactions take place
that change the chemical constitution of the polymer By hydrolysis and thermal
decomposition and oxidation of PET during its synthesis and melt spinning, carboxylic end groups are formed. Actually textile fibers and yarns contain
25–60 m. val/kg COOH groups, which is a remarkable quantity. Nevertheless such
fibers do not dye at all with basic dyestuffs. In addition to high crystallinity and
water repellency, the very low acidity of the carboxylic end groups of PET is responsible for this behavior.

An unsuccessful attempt to obtain this formula was made by substituting a certain amount of the terephthalic acid with trimesinic acid (1,3,5-benzene tricarbo-
xylic acid)[82–88] or by using carboxylated compounds having higher acidity due
to halogen atoms present in the α-position with respect to the carboxylic groups[84].

Disperse dyestuff with:	Affinity	
	high	low
Modified polyester fiber	A_m	B_m
Standard polyester fiber	A_0	B_0

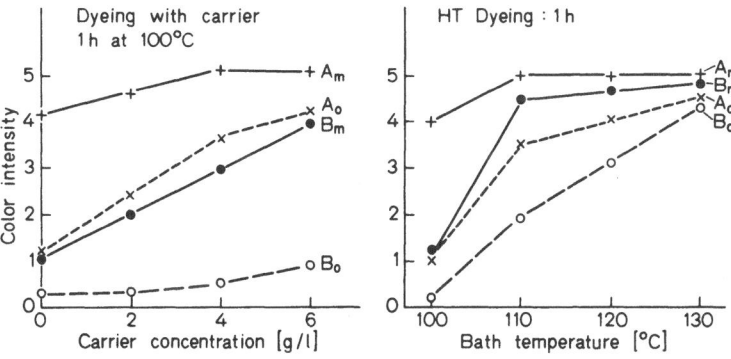

Fig. 9. Dyeing of texturized yarns of standard and modified polyesters with disperse dyes and
carriers or at increasing temperature and pressure (HT process)[81]

Much better results were achieved by a process based on the formation of co-polyesters containing strong acidic groups, especially sulfo- and to a lesser degree phosphoric groups. Since the number of papers and patents published on the subject is very large (not less than 60 in the last ten years) it is impossible to discuss them fully here. Table 6 contains a summary of some of the compounds mentioned for this purpose.

By adding 2–3 mol % of the sulfonated monomers or half of this amount, when the compound contains two sulfo-groups, to the reacting mass during the polycondensation stage, copolyesters are formed, which, after being transformed to yarns or fibers, can be satisfactorily dyed by basic dyestuffs and which also have a better uptake for disperse dyes. Blends with standard PET yarns are dyed with basic dye-stuffs, producing white reserves of the latter, while with disperse dyestuffs different shades are obtained, and when disperse and basic dyes are used in the same bath, two-colored effects are achieved. Methods for these dyeings and results obtained are reported in various papers[97–101].

The fastness of basic dyes on such copolyesters is generally satisfactory. Only the lightfastness is somewhat lower than normal, so that a careful selection of the dyestuffs is advisable. Attempts to improve the lightfastness were made in several ways. An interesting study in this regard was published by Maerov et al.[102] who after

Table 6.

Sulfonated benzene compounds

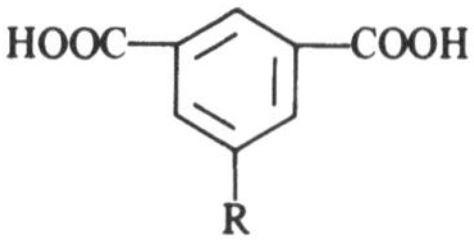

Where R is: $-SO_3Na$, $-R'-SO_3Na$, $-SO_2Na$, $-SO_2-O-C_6H_5$, $-SO_2-O-C_6H_4-SO_3Na$

85–90)

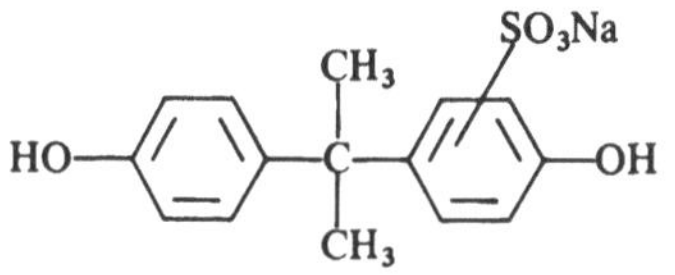

93–96)

Sulfonated compounds with condensed rings

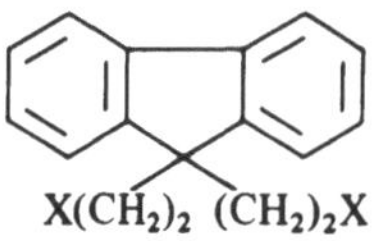

X=$-OH$,$-COOH$
$-SO_3Na$

91)

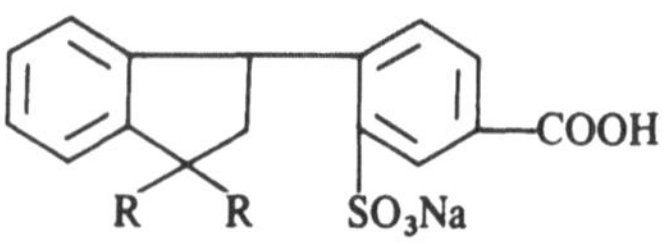

92)

Phosphoric compounds

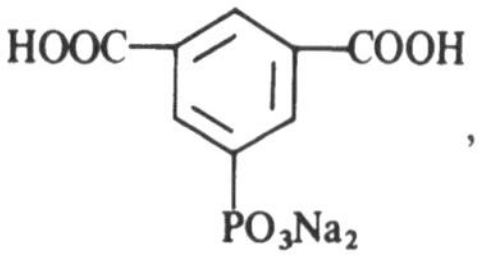

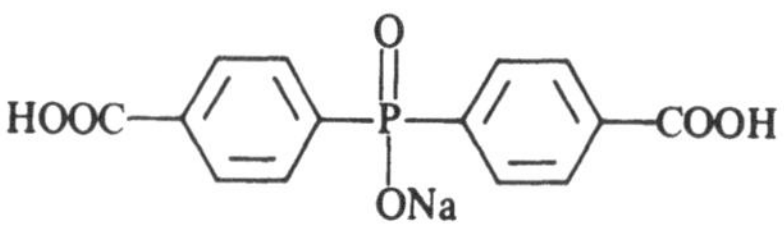

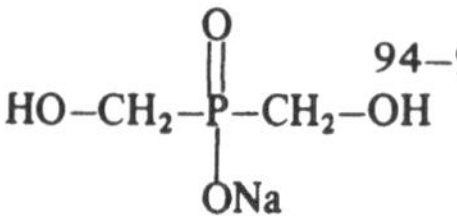

94–96)

measuring the most harmful light frequencies, investigated a method for making them harmless by applying products that have a screening effect and thus stabilize the color. One of these compounds is 2,2'-dihydroxy-4,4'-dimethoxybenzophenone ("Uvinil" produced by Gen. Aniline & Film Co,). It can be employed in the dye bath or by treating the textile before or after dyeing.

Copolyesters containing strong acid groups generally have a lower molecular weight and a higher melt viscosity than standard PET. They can also be used for the manufacture of low-pilling fibers (see Section 2.3).

Fibers produced from this kind of copolyester are more easily hydrolyzed than those from standard PET. All finishing treatments must thus be accomplished within a pH range of 4–9.

3.1.3 Dyeable Fibers with Anionic Dyestuffs

For making polyester fibers dyeable with acid dyestuffs it is obviously necessary to introduce basic groups in the polyester chain. This operation is not easy, amines having a comparatively low resistance to the elevated temperatures used for the polycondensation stage. Actually it was found that nitrogen-containing polyesters, prepared experimentally in the laboratory, darken more or less upon heating up to $270\,^\circ$C. Some improvement may be achieved by oxyalkylation of the amine, though complete thermal stability cannot be ensured by this method. Better conditions may be obtained by using amides or polyamides or amides with amine groups, eventually in the presence of suitable antioxidants. In this connection I.C.I. claims that good results may be obtained by adding to the batch 2,2'-dihydroxy-3,3'-bis-(2-methylcyclohexyl)-5,5'-dimethyldiphenylmethane[112, 113]. To avoid long heating times, the nitrogen-containing compound can be added directly to the chips during spinning.

Table 7 contains a list of compounds (amines and amides) that are claimed by patents for this purpose. This classification is partly fictitious, since amines, when

Table 7.

Aliphatic and aromatic amines	Heterocyclic compounds
Hexamethylenediamine[103]	Substituted piperazone[107]
Tribenzylamine[104, 105]	Substituted pyrrolidine[108]
Oxy-ethylated butylamine[106]	Substituted piperidine[109]
Oxy-ethylated aniline	Triazols[110]
	Substituted benzimidazols
	Polyvinyl pyridine[119]

Amides
Polycaprolactame = Nylon 6
Poly(hexamethylenediamineadipamide) = Nylon 6.6
Copolyamides from Nylon 6 and 6.6[111–118]

added during polycondensation, are apt to form amides by a reaction with free carboxylic groups present in the batch. From this point of view heterocyclic rings containing nitrogen (piperidine, piperazine, triazol, etc.) are more stable.

For thermal reasons, it certainly should be more suitable to add the amines or
amides directly to the chips during melt spinning, but results achieved in this way
are disappointing because of the disuniformity of color that is obtained with acid
dyestuffs.

It must be acknowledged that for the time being it is not possible to manufacture an acid-dyeable polyester fiber, despite the great interest in dyeing such fibers
together with wool.

3.1.4 Dyeable Fibers with Chelating Agents

It has been discovered that some azoic dyes, e.g., naphthylazochlorophenol,
naphthylazobenzothioazolyl-pyrrazolone, etc. show a chelating action by forming
complexes with bivalent metals, e.g., with nickel. This finding has induced some
manufacturers of synthetic fibers[120, 121] to add to molten PET heat-resistant, soluble or easily dispersable nickel compounds, e.g. Ni-dioctylphosphate, Ni-stearate,
etc. When 1—3 weight % of these compounds is added to the chips prior to spinning,
fibers are obtained that can be dyed satisfactorily with former azoic dyestuffs, the
color having a good fastness to light, washing, etc.

3.2 Antistatic Fibers

Without going into detail on the formation of static during the processing and use of
PET fibers, which has already been discussed by various authors[122, 123] it is useful,
for the sake of a better understanding, to bear in mind the following fundamental
concepts.

1. Static charges are generated when moving fibers or yarns are detached from
the contacting surface.

2. Static is dissipated toward air when the field force, due to the build-up of
charges on the fiber surface, exceeds the dielectric force of air. This depends on pressure, humidity, and the state of ionization of the surrounding air.

3. A stream of electrostatic (e.s.) charges flows along the yarn and consists of
a component (I_t) in the direction of the movement of the yarn

$$I_t = v \cdot \sigma (x)$$

where v = speed of the yarn (cm/s); $\sigma (x)$ = density of surface charges (Coul/cm at
the point x of the yarn or Coul/cm^2 when the value applies to plane textile constructions), a contrary component (I_r), which depends respectively, on the conductance,
on the specific surface resistance of the yarn ρ_F [Ω] and on the component (at the
point x) of the field force E_x in the direction contrary to the movement

$$I_r = \frac{E_x (x)}{\rho_F}$$

At equilibrium, therefore,

$$I_\infty = I_t - I_r = \sigma(x) \cdot v - \frac{E_x(x)}{\rho_F} = \sigma_\infty \cdot v$$

Supposing that E_x is proportional to σ_∞, i.e., to the density of surface charges at equilibrium, than E_x is given by $E_x = a(x) \cdot \sigma_\infty/\epsilon_0$, and at the separation point of yarn from the contacting surface $(x = x_0)$ charges formed by friction are

$$\sigma_0 \cdot v - \frac{a_0}{\epsilon_0 \cdot \rho_F} \cdot \sigma_\infty = \sigma_\infty \cdot v$$

where $a_0 = $ a factor dependent on the form and type of material, whose value, according to Löbel[123], is the range 0.2–5; $\epsilon_0 = $ dielectric constant, about 10^{-13} [Coul/Volt · cm]; $\sigma_0 = $ density of e.s. charges immediately after the separation point; this density can rise to a maximum of 10^{-9} Coul/cm, which is the dielectric value of air. Finally, the density of e.s. charges at equilibrium is

$$\sigma_\infty = \frac{\sigma_0}{1 + \dfrac{a_0}{\epsilon_0 \cdot \rho_F \cdot v}}$$

According to the investigations of several authors, σ_0 is in the range 10^{-9}–10^{-11} [Coul/cm]. If $a_0 = 0.2$, the product $\rho_F \cdot v$, sufficient to prevent static-related troubles, is

$$\rho_{Fl} \cdot v = 2.10^{-11} \; [\Omega \cdot cm^2/s] = 200 \; [G\Omega \cdot cm^2/s]$$

where $\rho_{Fl} = $ limiting surface resistivity and $G\Omega = 10^9 \; \Omega$.

It is known from experience that if the former product rises to about ten times the original value, troubles are relatively small, but they increase considerably when this limit is surpassed. For a yarn moving at a speed of 100 cm/s, the limiting surface resistivity is $\rho_{Fl} = 2 \; [G\Omega \cdot cm]$.

Some values of the surface resistivity of different textile materials at 20 °C and 65% r.h. are listed in Table 8[124]:

Table 8.

Material	Surface resistivity [log ($\Omega \cdot$ cm)]
Rayon	7–10
Cotton	7–11
Cellulose acetate	12–13
Polyacrilonitrile	12–13
Polyester	14–15
Polyvinyl compounds	15–16

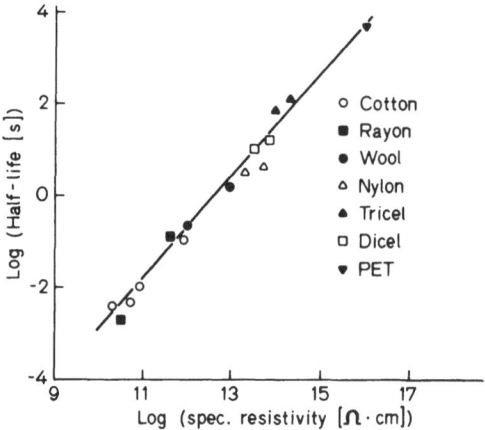

Fig. 10. Half-life period vs. specific resistivity of the fiber[134]

Surface resistivity of plane constructions (fabrics, rugs, etc.) is generally measured with ring-shaped electrodes[125]. In the case of loose fibers, the measurement is performed in cells between flat electrodes at a distance of 2 cm, the filling degree being 20%. From the volumetric resistivity ρ_V, relating to the cross section between the electrodes, the longitudinal resistivity ρ_G is calculated. This value can also be measured directly on a continuous filament and spun yarns between clamp electrodes.

The product of the limiting longitudinal resistivity ρ_{GI} and the speed of yarn, smooth processing considered, is $\rho_{GI} \cdot v = 2000 \; [G\Omega \; cm^2/s]$.

Frequently in addition to surface resistivity, the e.s. tension developed by friction under given conditions, or the half-life period of the charges, using, e.g., Schwenkhagen's device[126−131], is measured. The half-life period is linked to the resistivity of the materials by an exponential function, as shown by the data listed in Table 9[132−134].

Experimental values recorded on different fibers confirm this relationship (Fig. 10).

Antistatic treatments of polyester fibers were proposed in order to realize the following objectives:

A. Easier processing of yarns and fibers in all steps of the textile-manufacturing procedure. In this case the treatment can be restricted to the addition of antistatic compounds to the spin-finishing bath. These compounds do not necessarily have to be wash-proof.

Table 9.

Spec. resistivity	Half-life period
$(G\Omega \cdot cm)$	(s)
10^7	1100
10^5	11
10^3	0.11
10	0.0011

B. Lasting, wash-proof antistatic behavior, achieved by incorporation of antistatic compounds in the PET molecule in order to prevent the formation of e.s. charges by friction, especially in underwear or furnishing fabrics (curtains and carpets). In this way, rapid soiling of the textiles formed from PET should also be avoided, which sometimes is imputed to the e.s. charges on the fiber and to the electric field generated by them[135, 136]. This opinion is actually still under discussion, some authors[137] having shown experimentally the nonexistence of a relationship between the tendency of the fiber to build up e.s. charges and the speed of soiling of textiles. Nevertheless many fiber manufacturers go on heralding antistatic polyesters as soil resistant.

A. The *addition* of an *antistatic compound* to the finishing bath is common practice in processing PET fibers. Antistatic compounds used for this purpose consist primarily of *oxyethylated* fatty acids, amides, or amines[138, 139], which are all strongly antistatic and lubricating agents. The importance of the *degree* of oxyethylation of fatty acids for the resistivity and for e.s. charges, developed by friction of the fiber after treatment with such compounds, has been studied[140]. Specific resistivity of the yarn − for equal amounts of finish on the filament − decreases quickly to a level corresponding to 18 units of ethylene oxide, where there is a neat stop. The static formed by friction continues, however, to diminish to about 22 moles of ethylene oxide per mol of fatty acid, the decrease from 18 to 22 moles ethylene oxide being particularly sharp. Two different mechanisms thus seem to be responsible for the antistatic action of the oxyethylated compounds. Up to 18 moles of ethylene oxide, the decrease of resistivity prevails, and runs parallel to the moisture regain of the treated fiber. The lower e.s. charges generated by friction seem to depend on a structural change in the polyoxyethylene chain, which for a degree of polymerization of more than 18 moles of ethylene oxide, changes from a zig-zag line to one that meanders. The mechanism of action of these compounds, though not yet exactly known, indicates a degree of oxyethylation of about 22, reflecting the optimum of their antistatic effect.

Another group of high antistatic compounds is the *phosphoric* esters, especially those endowed with capillar activity, e.g., lauryl phosphate[141, 142], which are extensively, employed despite their toxicity. The antistatic effect of polyoxyethylated and phosphoric compounds is combined in the phosphoric esters of polyethyleneglycols[137]. Generally, the antistatic component of the finishing bath consists of a mixture of several compounds, in order to achieve the highest antistaticity with the smallest amount of these products.

Coating the filaments with such compounds considerably lowers the resistivity of the fibers and increases the maximum speed of the yarn, the product $\rho_G \cdot v$ remaining constant at about 2000 (see p. 22). The data in Table 10 may help to explain this behavior.

B. *Bulk antistaticity*. To achieve antistaticity that is resistant to washing, compounds are used that theoretically are like those added to the finish, i.e., they contain oxyethylene or phosphoric groups. Other compounds that can produce a similar effect are those containing ionic groups and even salts, e.g., KCl or LiCl[144]. However, the majority of patents claim, as antistatic agents, polyethylene glycols of different molecular weights or diamines, formed from polyethylene glycols by cyano-

Table 10. Continuous-filament PET yarn, 150 dtex/48 filaments

Finish	% of yarn	ρ_G [G$\Omega \cdot$ cm]	v_{max} [cm/s]	$\rho_G \cdot v$
Type A	0.17	260	8	2080
Type A	0.90	21	100	2100
Type B	0.15	440	5	2200
Type B	1.70	21	100	2100

ethylation and hydrogenation, which are condensed with adipic acid or better still with aromatic sulfo-compounds (5-sulfo-isophthalic acid). Actually the modifiers closely resemble those used for making polyester fibers dyeable with disperse or cationic dyestuffs.

A short list of co-monomers suggested in recent patents is reported in Table 11 together with their antistatic effects expressed by the drop of resistivity, the decrease of e.s. charges due to friction and that of the half life period of the same. For lack of information on the test methods, the comparison of the data is not possible, but they are nevertheless useful for evaluating the effect of the single compounds.

The process of including these compounds in PET varies according to the thermal stability of the antistatic agent. It may be added during polycondensation or by mixing the chips with the modifier, dissolved or dispersed in a medium eventually capable of swelling the PET and which afterward will be removed by vacuum evaporation in a rotating dryer. As stated previously, the amount of modifier added should be as low as possible in order to avoid losses in physical and textile properties of the fiber.

3.3 Flame-Proof Fibers

Until a few years ago, papers and patents regarding flame-proof polyester fibers were rare, probably because these fibers, even, when consisting of standard PET and not coated with fireproof products, are fairly flame resistant. Data obtained under strictly controlled conditions, based on a measurement of the "limiting oxygen index" (LOI), i.e., the lowest oxygen concentration necessary for maintaining regular combustion of the material, have been published[161−164] and are listed in Table 12.

The figures in Table 12 depend in part on the textile construction of the sample and on the denier and the length of the fiber.

Other test methods have been proposed, which are less exact than measurement of LOI, but which essentially reflect practical conditions. Some of these methods have been officially adopted in several countries[161−165].

In the *U.S.A.*, the old method AATCC 34 − 1952, has been substituted by a newer one, Doc − FF − 1 − 70, which has been made compulsory. The method consists of ignition of the horizontally placed sample by means of a methenamine tablet. The flame resistance is judged positive when the burnt circular area has a diameter of not more than 6 inches = 15.24 cm.

Table 11.

Co-monomers	Resistivity (GΩ · cm)		e.s. charges (V)		Half life (s)		Ref.
	without	with	without	with	without	with	
1.5% Polyethylene glycol + 1% nonylphenoxypropane-potass.sulfonate	—	—	4500	450	100	4	145, 146)
3% Polyethylene glycol (mol. wt. 20,000)	—	0.1	—	—	—	—	147)
1% Cond.prod.alkylene-oxide with mono- or diamine	—	—	2100	200	—	—	148)
The same with adip.acid or sulfo-isophthal. and Nylon 6	—	—	3850	150	—	—	149, 150)
10% Polyvinyl-alcohol	10^6	10^3	—	—	—	—	151)
0.1% 2-Naphthol(3,6-sodium-disulfonate)	—	—	960	36	—	—	152)
4% Co-pol. PEG + diamine + Nylon 6	40	0.23	—	—	—	—	153)
5% $H(OC_2H_4)_m-O_3S-\phi-SO_3-(C_2H_4O)_n \cdot H$ (m + n = 45)	9000	0.36	—	—	—	—	154)
5% $[(-OC_2H_4)_{70}-O-P\overset{O}{\diagdown}\underset{Ph}{}]_n$	9000	0.23	—	—	—	—	155)
10% Polyethylene glycol (mol. wt. = 8200)	—	—	4900	550	—	—	156)
5% $HO(C_2H_4O)_6-(CH_2)_3-SO_3Na$	150	8	(moist. regain)		0.3 →	1.2%	157)
1% Adduct of Triazine and PEG (mol. wt. 3000)	—	—	—	—	500	5	158)
3% $(C_6H_5)_2N-COO(C_2H_4O)_{10}H \cdot H_3PO_4$	1000	1	—	—	600	5	159)
0.5% Sodium glycerophosphate	380	28	—	—	—	—	160)

Table 12.

Fiber	Limiting oxygen concentration (vol)
Cellulosics	0.18 −0.20
Polyacrylonitriles	0.20 −0.21
Polyesters	0.25 −0.30
Polyamides (Nylon 6 and 6.6)	0.275−0.30
Wool	0.287−0.30
Modacrylics	0.30 −0.35
Aromatic polyamides (Nomex)	0.37 −0.41
Polyvinyl chlorides	about 0.45

In *Germany* the test methods DIN 53906 and 54332 (1971) are used. They are based on an evaluation of the burnt area of the vertically placed sample, the ignition time by a gas burner depending on the fabric weight. The DIN 53907 method is performed with a sample placed horizontally in a beaker of given dimensions, and ignited with a flame of 0.3 ml ethyl alcohol (96%).

Correlation of the results obtained by these methods with the LOI shows that fibers having an LOI value of 0.25 or less are also considered flame proof by the U.S.A. and DIN tests.

At any rate, PET fibers are found to be comparatively fire resistant, surpassing cellulosics and acrylics, but not filaments formed by polyvinylchlorides. The later, however, though not burning, give off, in the case of fire, poisonous hydrochloride vapors.

The situation changes completely when considering *blends* from polyester fibers and cellulosics, which actually form the greatest part of the fabrics for underwear and light suits. It has been found that these kinds of fabrics have lower flame resistance than both of the components forming the blend[166]. As shown in Figure 11, the LOI value of the most frequently used polyester-cotton blends for example 65/35 or 50/50, is lower than that of either cotton or polyester. This circumstance has probably been the reason why so many suggestions have been made for increasing the fire resistance of the blends up to the safety limits established by law.

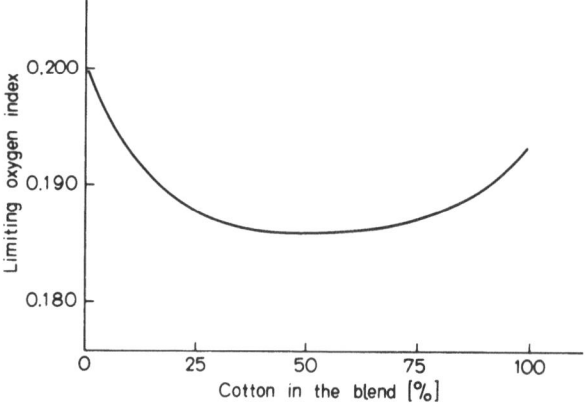

Fig. 11. Flame resistance of blends from PET fibers and cotton as measured by the limiting oxygen index[166]

Processes with this objective, like those used for improving antistaticity of fibers, may be divided in two groups[167]:

A. Coating of single fibers or preferably of blends with fireproof products.

B. Embodying compounds in PET capable of improving the flame resistance of the polyester.

A. The products used for this purpose are the same as those employed for the flame-proofing of cellulosics. Some of them are listed in Table 13.

It is important to keep in mind that considerable amounts of these compounds (about 20% of the fiber weight) are required in order to develop an appreciable flame resistance and that a final thermal treatment is necessary to make the coating sufficiently wash-proof. The use of these compounds results in tenacity loss of 25%–30%. Another difficulty arises when treating blends of standard PET fibers and cellulosics with finishing baths containing such compounds, which generally have a higher affinity for cellulosics than for polyesters. This trouble may be partially, overcome by using finishing baths of a suitable composition[170–175].

Actually, the use of these treatments permits the safety limits established by United States regulations for upholstery fabrics (curtains, rugs, etc.) to be attained. However, after several washings, as e.g., in the case of nightgowns, some of the resistance has been lost, so these values can be accepted only with reservations.

B. Compounds that improve fire resistance when introduced into the PET chain, are essentially of the same type as those described in Section A. The most frequently used are the phosphorus compounds, known for their good compatibility with PET, being employed normally for thermal stabilization of polyesters. Other compounds mentioned in this connection contain nitrogen or bromine (Table 14).

Table 13.

Tetrakis-hydroxymethyl-phosphonium chloride $[P \equiv (CH_2OH)_4]Cl$	168, 169)
Tris-aziridinyl-phosphonium oxide $O = P \equiv (N < C_2H_4)_3$	170)
Tris-hydroxymethyl-melamine	171)
Tris-hydroxymethly-melamine-phosphate	172)
Tris-(2,3-dibromo-propyl)-phosphate ⎫	173)
Tris-(2-bromo-ethyl)-phosphate ⎬	
Bis-(hydroxymethyl)-ammonium-phosphate	174)

Table 14.

2-Oxo-2-phenyl-1,3,2-dioxaphosphorane	176)
Tris-(hydroxyethyl)-phosphite	177)
Poly(3,3-bis-(bromoethyl)-ethylene glycol $HO[C(C_2H_4Br)_2CH_2-O]_nH$	178)
Di-ethyl-3,4-di-bromoadipate	179)
1,2-dihydroxyethane-1,1-diaminodiphosphate	180)
Triphenylphosphate + hexabromobenzene	181)
Trimethylphosphate + polyethylene glycol	182)
Triphenylphosphate + polycaprolactame	183)

By introducing 2–20 weight % of these compounds in the PET molecule an increase of up to about 40% of the LOI value may be achieved.

The fire resistance of polyester fibers can also be improved by substituting part or all of the ethylene glycol employed by 1,4-butandiol and introducing at the same time in the polyester chain phosphorus or bromine-containing compounds[184–186].

It should be recalled here that some time ago a highly fire-resistant fiber, named "Enkatherm" was produced by polycondensation of terephthaloyl chloride and oxaminohydrazone (a compound readily formed from cyanogen and hydrazine), according to the reaction:

$$n\ Cl-CO-\phi-CO-Cl\ +\ n\quad \begin{array}{c} H_2N-N \quad N-NH_2 \\ \diagdown C-C \diagup \\ H_2N \quad\quad NH_2 \end{array} \longrightarrow$$

$$\left[\begin{array}{c} NH-N \quad N-NH \\ -\phi-C \quad C-C \quad C- \\ \parallel \quad\quad \parallel \\ O\ H_2N \quad NH_2\ O \end{array} \right]_n \underset{H^+}{\overset{OH^-}{\rightleftharpoons}} \left[\begin{array}{c} N-N \quad N-N \\ -\phi-C \quad C-C \quad C- \\ O^-\ H_2N \quad NH_2\ ^-O \end{array} \right]_n$$

This polymer in its tautomeric form is soluble in caustic alkali and can be spun in normal rayon spin-baths. The most interesting fact is that salts of bivalent metals (Zn^{2+}, Fe^{2+}, etc.) added to the spin-bath react rapidly with the polymer, forming chelates. Such fibers are highly flame proof, their LOI value being greater than 0.50, i.e., even higher than the corresponding value of the "Nomex" fiber and of polybenzimidazoles, which are used particularly in aeronautic construction.

3.4 Soil-Proof and Soil-Releasing Fibers

To understand the problem of making polyester fibers more resistant to soiling and to prevent the redeposition of soil from detergent baths, it is necessary to discuss the phenomena of soiling and soil removal. This discussion is complicated, however, by the different mechanisms involved in these phenomena. The problem is made still more complex because the so-called soil consists of heterogeneous matter formed by many components having different affinities for fibers, which can also change as a result of the combined action of the components or due to the construction of the textile goods (yarns, fabrics). For this reason, an exact description of the physics of soiling and washing is rather difficult. In fact, at present, there is no satisfactory theory to explain mechanism of these phenomena and, consequently, to improve the resistance of fibers to soiling and soil release.

A short summary of the different ideas expressed on the subject in the vast literature shall be given here in order to clarify the fundamental concepts.

Soiling of textiles takes place essentially in three ways:

1. by contact with *solids,* for example (a) by absorption of grease particles from human skin (sebum) by underwear and bed linen; (b) by friction of upholstery fabrics, outer wear, and tableclothes on different objects; (c) by walking on carpets with shoes.

2. by contact with *liquids:* The tendency of polyester garments to absorb dirt particles from aqueous detergent systems in laundering is well known.

3. by contact with *air:* Dust particles suspended in air tend to be deposited on the free surface of textile goods such as upholstery and curtain fabrics.

Without going into details on the physics of these phenomena, we shall discuss the behavior of PET fiber and of its blends with other fibers, specially with cellulosics, in this regard.

1. The most remarkable and visible effect of this behavior results in the appearance of linen after a short period of wear. It is well known to housewives that shirt collars and cuffs made from PET/cotton blends show darker edges than those of shirts made from pure cotton. There is, however, a lack of reliable quantitative data on the subject and the few available data are contradictory. For instance, by extraction with suitable solvents after 24 h of wear, the cotton and the PET/cotton 65/35 shirts, contained the following amount of fatty matter[187]:

	% Medium	% Minimum	% Maximum
Pure cotton	1.17	0.68	1.76
PET/cotton 65/35	1.09	0.66	1.74

There is almost no significant difference between the two series of values. A reasonable explanation for these data, in contrast with the appearance of the garments, may be the fact that dirt (grease and pigments) absorbed by friction stands out much more on PET fibers with round cross section than on cotton with flattened and twisted surface.

The uncertainty of the results has discouraged investigators from using this method, based on the friction of fabrics with solids, for measuring the soiling behavior of fibers.

2. Treatment with aqueous detergent systems containing different kinds of dirt particles in suspension or in emulsion is the most frequently used method for investigating the soiling tendency of fibers and of textile goods.

Two variables — leaving aside fiber quality and construction of the textile — are essential for evaluation of the redeposition behavior of dirt[188-195]

a)   the kind of soil experimented with,

b)   the test method for measuring the amount of absorbed particles.

a) To reproduce practical conditions as exactly as possible, most substances investigators employ:

a mixture of fatty matters having the same or similar composition, like sebum,

a mixture of pigments formed by clay particles or, preferentially, by air dust collected conveniently from the filters of a conditioning plant.

b) To measure the quantity of absorbed soil, the extraction method may be used. With this method, however, only the fatty matter is recorded. The photometric method is refrequently used, that is the measure of reflectance at a given wavelength. When this test is used on the original fabric and after soiling and washing, the change of reflectance, i.e., the loss of whiteness of the samples, subject to many cycles of soilings and washings, is measured.

It is difficult to compare the results since authors use different methods for computing them. For example Bowers[187] expresses the value of absorbed soil before washing (BW) and that left after washing (AW) by the formulas

$$BW = \frac{R_0 - R_s}{R_0} \cdot 100 \qquad AW = \frac{R_0 - R_w}{R_0} \cdot 100$$

where R_0 = reflectance of the original fabric; R_s = reflectance of the soiled fabric; and R_w = reflectance of the soiled fabric after washing.

Another way to express the results is the method used by Byrne[195] by calculating the soil concentration on the sample before (SAD) and after washing (GAD)

$$SAD = \log \frac{R_0}{R_s} \qquad GAD = \log \frac{R_0}{R_w}$$

Sontag[191] gives the results of his tests after transformation of the reflectance in K/S values, by applying the equation of Kubelka-Munk[196]

$$K/S = \frac{(1 - R_f)^2}{2 R_f}$$

where K = absorbed light coefficient; S = diffuse light coefficient; and R_f = measured reflectance.

On the whole, photometric results computed by the various formulas agree at least in that they show a similar trend.

Quite different values are obtained by a new technique based on the inclusion of organic compounds containing isotopes (^{14}C or ^{3}H) in the soil and the measurement of the radioactivity of the sample before and after washing with a suitable counter[197]. The different figures registered with this procedure as compared to the photometric method may depend on the fact that by the latter method the surface soil concentration is recorded while the measure of radioactivity gives the total amount of absorbed soil.

A calorimetric test[198] based on a measurement of the heat evolved when treating more or less stained fibers with 37% caustic soda, does not seem sufficiently sensitive for recording the amount of soil on the fiber.

At any rate, using these studies, especially employing photometric methods, the following facts were ascertained to be in satisfactory agreement between the single statements:

there exists a strict correlation (see Fig. 12) between soiling and hydroophylic tendency (expressed as moisture regained at 23 °C and 65% r.h.) a more hydrophylic fiber having a lesser tendency to absorb fatty matter and soil.

Particularly on PET fibers, cotton, and various blends, the data in Table 15 on soil uptake (SAD) and soil left after ten cycles of soiling and washing (GAD) were recorded by Byrne[195] with dirt formed from fatty matter and dust or only from dust.

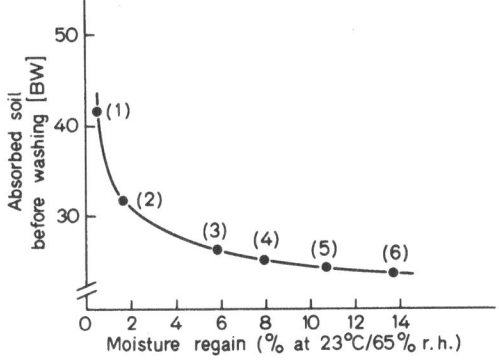

Fig. 12. Soiling of fibers as a function of their moisture regain (at 23 °C/ 65% r.h.)[187]: (1) PET; (2) poly-acrylics; (3) cell. acetate; (4) cotton; (5) rayon; (6) wool

Table 15.

Fiber	Fatty matter and dust		Dust	
	SAD · 10^3	GAD · 10^3	SAD · 10^3	GAD · 10^3
PET	390	52	29	17
PET/cotton 67/33	170	76	44	30
Cotton	18	10	23	13

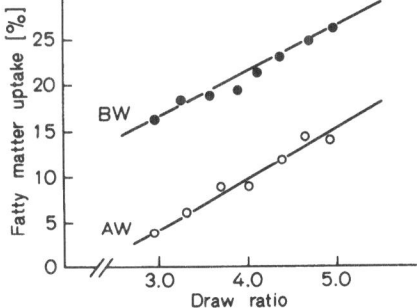

Fig. 13. Soiling of PET fibers as a function of stretch

PET fibers evidently have a greater affinity for fatty matter and a lower one for solid particles, while the contrary is true for cotton. This behavior may be explained by:

the higher zeta potential of PET fibers in comparison to cotton [199] for example Z_{max} = 61 mV vs. 23 mV

the similarity between the disperse dye uptake and soil redeposition.

It could be argued that the latter phenomenon is influenced by the molecular structure of PET fibers — crystallinity and orientation — which, as is well known, depend on stretch and heat-setting. In fact this correlation exists, but is not very clear since with higher stretch the fatty matter uptake increases both before and after washing (see Fig. 13), which is the opposite of dyeability; with higher temperature of heat-setting and ensuing higher crystallinity the grade of soiling of PET fibers does not change appreciably (Table 16):

Table 16.

Temp. of heat-setting °C	Crystallinity %	% Fatty soil uptake	
		BW	AW
50	37	58.2	34.0
100	42	58.3	34.4
150	43	60.5	33.0
200	50	59.7	33.2
225	53	55.0	28.0

Another factor that has considerable influence on soiling is related to the finishing treatments, the so-called permanent press and wash-and-wear finishes, which furthermore improve the excellent performance of the garments. These treatments are based on the anticrease concept and cause cross-linking of cellulosics, due to the formaldehyde or glyoxal content of the finishes. In Table 17 the soil uptake before and after finishing is listed together with the moisture regain of the fibers.

Table 17.

Fiber	% Soil uptake BW	% Moisture regain at 23 °C/65% r.h.
Cotton	28.4	7.3
Cotton after finish	37.1	4.9
PET/Cotton blend 50/50	34.0	3.8
PET/Cotton blend 50/50 after fin.	44.0	2.8
PET	56.0	0.5
PET after finish	50.0	1.2

Such finishes increase the soiling tendency, especially of blends, and they have been the starting point for the studies aimed at preventing the problems caused by them.

All these facts indicate that it is the surface of the filaments that is the seat of their tendency to soil and their washability. Therefore, to improve this characteristic, it is necessary to influence the surface. Indeed the majority of the proposed processes do not consider the introduction of a particular compound into the PET molecule, capable of changing its properties, but rather suggest the use of special finishes, whose purpose it is to link the coating to the filament in order to make it resistant to wash and wear.

The products described in the literature can be divided into three classes.
A. *Hydrophilic* polymers or copolymers, or processes designed to modify the fiber surface, making it hydrophilic by chemical reaction,
B. *Acrylic* polymer or copolymers,
C. *Perfluorated* compounds or polymers.

A. Any increase of the hydrophilic properties of polyester fibers should improve their resistance to soiling by fatty matter and facilitate their removal by washing. Because of the great number of existing papers and patens related to the subject, only the basic concepts shall be considered here[200-212].

To coat PET fibers and its blends with a strongly hydrophilic layer, the majority of papers and patents suggest the use of polyethylene glycol, the only variables being the methods for a durable linking of this polymer to the fiber surface.

For example I.C.I.[208] claims a treatment with a product named Permalose TG or Cirrasol PT, formed by the reaction of terephthalic acid, polyethylene glycol, and ethylene oxide. This product can also be employed in solutions of dimethylsulfoxide, which causes PET to swell. A final thermal treatment at 180–190 °C for 30 s serves to fix the layer to the surface. This process may be applied both to raw fibers and to fabrics, finished by a permanent press treatment, or in the same bath containing the aminic resins used for this treatment.

The Teijin Co.[212] describes the processing of polyester fabrics by a block-co-polymer formed by polyethylene glycol (mol. wt. about 800) and PET in dioxane solution. After evaporation of the solvent and final heat-setting, the wetting time of the fabric sinks from about 180 s to 6 s or less.

Atlas Chem. Ind.[209] suggests fixing of polyethylene glycol (mol. wt. about 6000) to the PET fiber surface through an urethane bridge, which ensures good resistance of the hydrophylic layer to washing.

The introduction of hydrophylic compounds during melt spinning is claimed in a patent of the Monsanto Co.[202] by adding pellets of polyethylene glycol (mol. wt. about 9000) -dibenzoate or -diacetate to the PET chips. The Hoechst Co. suggests the incorporation in the same way of 1%–4% 4-amino-N-stearylbutirylamide: $H_2N-(CH_2)_3-CO-NH-C_{18}H_{37}$.

Toray Ltd.[210] claims a remarkable improvement in the resistance to soiling by treating PET fibers containing polyethylene glycol or aromatic sulfo-compounds with a 1% caustic bath.

In this connection a still more drastic treatment with a 10% caustic solution (60 °C for 10 min.) of PET-cotton blends 65/35 and 50/50 has been suggested[213], by which the washing performance can be considerably improved though at the expense of the mechanical strength of the fabric.

By coating fibers with 2%–10% of previously mentioned finishes, soiling of fabrics from standard PET or from blends with cellulosics can be reduced, and the negative effect of the permanent press treatment can be counterbalanced.

B. From a theoretical standpoint finishing with *acrylates* means coating the fibers with a layer of a polymer with carboxylic groups pointing outward[214, 215]. Owing to the polar and hydrophilic nature of these groups and their negative charge, conditions are created that are apt to improve resistance to soiling and to facilitate soil removal. The use of another carboxylated polymer i.e., carboxymethylcellulose (CMC), which is used primarily for finishing fabrics from cotton and its blends, is based on the same concept. In both cases dirt particles are deposited outside of the carboxylic groups and not on the fiber, thus facilitating soil removal.

In practice the CMC finish does not noticeably improve the resistance to soiling and the launderability of fabrics because of the insufficient bond between CMC and the fiber. However, by applying certain special conditions, i.e., low-viscosity CMC together with dimethyloldihydroxyethylene-urea (instead of dimethylolethylene-urea) and $MgCl_2$ as catalyst, a satisfactory washability of fabrics finished in this way is achieved[214].

Opinions on the effectiveness of acrylates are contradictory, probably owing to the great number of possible combinations between acrylates, methacrylates, and other vinyl compounds, some of which give rise to high performance while others are less effective. Among the many papers on the subject, the following illustrate the attempts to promote the antisoil effect of acrylic copolymers[216-221].

The Deering Milliken Co.[218] suggests addition of 10% of an emulsion containing 25% of a copolymer formed from ethylacrylate and acrylic acid, or the same together with N-methylolacrylamide. After drying and heat-setting the washability index (min. = 1, max. = 5) rises to 3.5—4 after 10 launderings, the same without finish being 1. In another example the heat-setting is replaced by irradiation of about 2 megarads, achieving grafting of the acrylic component.

Du Pont[216] advises the use of polyacrylic acid cross-linked with 3% pentaerathrol.

In a patent of BASF[219] it is claimed that by addition of a mixture of polyacrylic or copolyacrylic-methacrylic acid and ethyl-orthosilicate to the permanent press bath a very good soil proofing is achieved, which, however, does not seem to be resistant to washings (see Kleber[194]).

According to a patent of the Roehm & Haas Co.[220] fabrics from PET/cotton blends are treated with a 5.5% suspension of copolyacrylic-methacrylic ethyl, i-butyl, or lauryl ester in trichloroethane. After evaporation of the solvent and heat-setting at 160 °C for 2 min, a satisfactory resistance to soiling is observed, also after several washings.

The Monsanto Co.[221] claims treating fabrics with a mixture of a permanent press finish and vinylacetate-maleic anhydride copolymer. After heat-setting at 160 °C for 2 min, the washability index is 3.6, the same index without copolymer being 1.0.

From the extensive research work of Peper and co-workers[188, 189], the importance of hardness of the acrylic layer becomes evident. If hardness is lower than about 60 degrees on the Durometer scale, even more soil is absorbed, and washing at 30 °C or more presents greater difficulty than without finish.

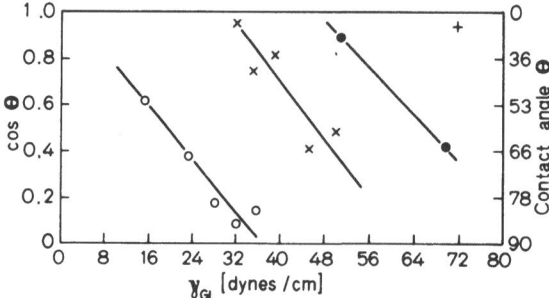

Fig. 14. Contact angle dependent on surface tension of the liquid phase γ_{Gl} on different materials[189]: + cotton, ● acrylic-methacrylic copolymers, x silicons, ○ perfluorated polymers

The rise of the negative zeta potential due to acrylates or to CMC should theoretically hinder the absorption of dirt particles, which generally have negative charges. There is no definite proof of this event; however, it is known that finishes having a positive charge (aminoacrylates, melamine resins, etc.) are decidedly harmful in this regard.

C. Owing to the nonpolar ends of the *perfluorated* compounds, they represent a model quite contrary to what is claimed in previous sections on soil-proof finishes. However, because of their low surface tension, they are at the same time water and oil repellent, i.e., the contact angle with both water and with oily matter is rather high.

It is well known that the contact angle θ between phases depends on the interfacial tensions

$$\cos \theta = \frac{\gamma_{GS} - \gamma_{SL}}{\gamma_{GL}}$$

where γ_{GS} = surface tension of gas-solid interface; γ_{SL} = surface tension of solid-liquid interface; and γ_{GL} = surface tension of gas-liquid interface.

As shown in Fig. 14 cotton is completely wetted   by pure water (γ_{GL} = 72 dynes/cm) while wetting of an acrylic layer is achieved only when the surface tension of the bath is less than 48 dynes/cm. The standard perfluorated finishes (fluoro-acrylates, etc.) are not wetted by aqueous solutions nor by oily matter with a surface tension of about 30 dynes/cm or less. However, when fabrics finished with these products are dipped in water and then a drop of hexadecane is introduced under the sample (compound which may be chosen as model for oily matter), it will spread quickly on the finished fabric, thus showing the preference of the perfluorated finisch for oil over water[222]. Moreover, if the fatty matter enters the fabric by friction it is difficult to remove it by washing from the perfluorated finish layer on the fiber.

In short, finishes formed from perfluorated compounds have not been successful in practice, despite their repellency to oily matter.

$$(CH_2-CH_2-HN-CH_2-CH_2-N-CH_2-CH_2-NH-)x$$
$$\underset{\underset{\underset{NH_2}{|}}{\underset{CH_2}{|}}}{CH_2}$$

$$+ \quad C_7F_{15}\overset{\overset{O}{\|}}{C}-O-C_2H_5$$

(A)   (B)

$$(CH_2-CH_2-HN-CH_2-CH_2-N-CH_2-CH_2-NH-)x$$
$$\underset{\underset{C_7F_{15}-\overset{\overset{O}{\|}}{C}-NH}{\underset{CH_2}{|}}}{CH_2}$$

- - - →

(C)

Attempts were then made by several investigators to unite in the same molecule hydrophilic and perfluorated groups, thus forming hybrid compounds, easy to wet with water but at the same time oil repellent and consequently soil resistant. Some papers[223] describe the synthesis of such a compound (C) by making polyethylene-imine (A) react with ethylperfluoro-octyl ester (B) according to the formula (see page 123).

By treating cotton and polyester blend fabrics with a solution of this compound in 2-ethoxyethanol ($C_2H_5 \cdot O \cdot C_2H_4 \cdot OH$) better results are obtained - even without heat-setting — than with standard perfluorated finishes, regarding the removal and the redeposition of fatty matter, but not the initial soil uptake. The latter fact is in agreement with the presence in the molecule of imine groups having positive electric charges that attract soil particles.

The overall performance of the finish is improved by a hybrid copolymer developed by the research department of the 3M Co.[222] and described by the following formula.

$$
H\left[\begin{array}{c} OCH_2CH_2NO_2SC_8F_{17} \\ | \\ C=O \\ | \\ CH-CH_2 \end{array}\right]_3 \left[\begin{array}{c} CH_3 \\ | \\ -S-CH_2CH-\overset{O}{\overset{\|}{C}}O-(CH_2CH_2O)_4-\overset{O}{\overset{\|}{C}}-CH-CH_2S- \\ | \\ CH_3 \end{array}\right]_{10} \left[\begin{array}{c} OCH_2CH_2NO_2SC_8F_{17} \\ | \\ C=O \\ | \\ CH_2-CH \end{array}\right]_3 H
$$

In this compound the oxyethylene chain represents the hydrophylic function, while the perfluorated groups linked to the acrylic chain are less frequent than in this type of standard finish.

Fabrics from polyester/cotton blends, processed after the permanent press treatment with this product and heat-set at 166 °C for 15 min, exhibit the values of oil repellency (ORP) and of launderability (L), as compared with those recorded on fabrics processed with finishes having only hydrophylic properties, shown in Table 18.

A hybrid finish of this type evidently has a very marked effect on the prevention of soiling and facilitation of soil removal. An important drawback, however, seems to be its high price.

Many other perfluorated compounds are recorded in patents, a list of which is given[224–231] without pretending to be complete. The products may be classified as standard (generally perfluorated acrylates) or hybrids, i.e., with highly hydrophilic groups in the molecule. Some patents mention the possibility of adding perfluorated compounds, e.g., $HOOC-\phi-C:(CF_3)_2-\phi-COOH$ to the batch in partial substitution of terephthalic acid used for the synthesis of PET[232, 233].

The importance of treatments for improving soil resistance and soil release properties has been extensively discussed lately, and it was questioned whether the

Table 18.

Fabric PET/cotton	Conc. in finish bath (%)	Hydrophylic finish origin. ORP	L	after 5 wash. ORP	L	Hybrid finish origin. ORP	L	after 5 wash. ORP	L
50/50	0.5	0	2	0	3	6	4.5	2	3.5
65/35	0.5	0	2	0	2.5	6	4.5	2	3.5
50/50	1.0	0	3	0	3.5	6	4	4	4
65/35	1.0	0	2.5	0	3	6	4.5	4	4
50/50	2.0	1	3	0	3	5	4.5	5	4
65/35	2.0	1	2.5	0	3	6	4.5	5	4.5
50/50	3.0	2	3	0	3.5	6	4.5	6	4
65/35	3.0	2	3	0	2.5	6	4.5	6	4.5

Classification: ORP (°) minimum = 0 maximum = 8
             L   minimum = 1 maximum = 5
(°) Method AATCC 118 · 1966 T [Am. Dyest. Rep. 56, P 112 (1967)]

unsatisfactory behavior of PET fibers is of the same consequence in Europe as it apparently is in the U.S.A. In fact the permanent press process has not been developed in Europe on a particularly wide scale, and the laundering difficulties encountered when washing at low temperature, as practiced in the United States, do not appear at higher temperatures[234]. Moreover, treatments with perfluorated compounds are expensive, and they diminish the mechanical resistance of the fabrics. Also in the U.S.A. the permanent press concept has been criticized recently[235]. While this process undoubtedly holds good for pure cotton garments, its value sinks in the case of fabrics made from PET fiber/cotton blends. The processes for improving soil resistance and soil release properties are apt to lose much of their actual interest, as they no longer employ the permanent press finish.

3. It has long been believed that the greater soil uptake by polyester fibers as compared to cellulosics, due to the contact with solids and liquids, should also occur upon *contact with air*, i.e., with aerosols formed by dust particles suspended in air. This supposition was based mainly on the fact that standard PET fibers tend to become charged with static electricity (see Sec. 3.2) and that this phenomenon should cause the deposition of dirt particles from air, e.g., similarly to the way that in electric precipitators, dust particles deposit on the electrode surface. Moreover, it is known that during spinning and weaving, fibers attract dust particles due to electrostatic charges. In support of the theory of soiling by aerosols is the fact that some antistatic finishes, especially oxyethylated compounds, show a marked antisoil effect. In the last few years this theory has been thoroughly investigated[236–241] on textile goods normally exposed to air.

In an extensive study, Reichle[236] elucidated these events by comparing the behavior of goods from synthetic fibers endowed with high resistivity (10^{13} $\Omega \cdot cm$) and a high tendency to build up electrostatic charges with that of cellulosics having low resistivity (10^9 $\Omega \cdot cm$). It was found that fabrics made from synthetic fibers charged by static of an increasing tension in the range of 500–5000 V (electric field at a distance of 5 cm = 70 to 680 V · cm) absorb dust particles from the air, the

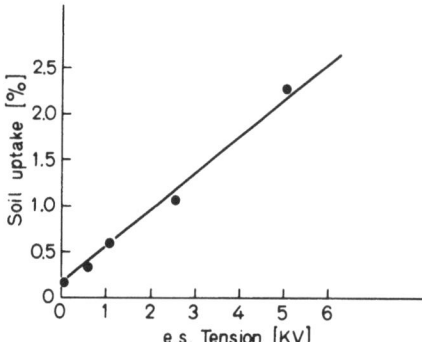

Fig. 15. Dust particles absorbed from air after 2500 h as a function of applied tension[236]

take-up being a linear function of the applied tension (Fig. 15). Soil concentration after 2500 h exposure to air is measured photometrically with monochromatic light, and the reflectance is transformed into soil units by the Kubelka-Munk equation. The experiments were then repeated by exposing for the same amount of time a curtain made from alternating stripes of synthetic yarns and of cellulosics, the curtain was moved several times every day in order to reproduce practical conditions. The results of the photometric measurements, made in the upper, middle, and lower zones of the curtain stripes, are listed in Table 19.

No significant difference of the amount of soil absorbed by the two materials under identical conditions was found. Apparently the different electric properties of the fibers do not influence their soiling tendency by aerosols.

The incongruency between the model experiment with variable electric tensions applied to the fabric and the practical conditions may be explained by the fact that the electric field measured at a distance of 5 cm under the latter conditions is very low, not surpassing $100 \, V \cdot cm$, so that the lack of absorption of dust particles appears to be justified by this circumstance.

Other authors express similar opinions on the soiling of fabrics by aerosols. Goodrich et al.[238] found, by investigating the behavior of curtains made from polyester, cotton, and blends, exposed to air containing different suspended particles (tobacco smoke, etc.), that PET fibers have the lowest soiling tendency.

From this point of view the use of antistatic finishes may seem senseless. However, it should be borne in mind that the conditions described are quite different from those existing during the processing of the fibers (spinning, weaving), so that the presence of antistatics in the lubricating mixture is not only useful, but quite necessary.

Table 19.

Yarn	Reflectance at $\lambda = 450 \, m\mu$					Soil concentration
	not exposed fabric	exposed fabric				
		upper	middle	low	medium	
Synthetic	68.7	58.1	58.1	58.0	58.1	0.80
Cellulosic	68.5	57.2	57.2	56.1	57.0	0.90

4 Conclusions

The performance of polyester fibers for certain end uses can be greatly improved by modification of the original chemical structure of the PET molecule. In this way it is possible to eliminate many of the short-comings that have hitherto slowed down or even hindered the use of these fibers in some textile areas.

Modifications, however, have to be made with moderation, so as the avoid worsening of the excellent physical properties of the standard PET fiber. The problem is still subject to research in order to find the best conditions for balancing the contrasting effects of modification and to achieve a substantial improvement of the behavior of the fiber, while retaining its traditional advantages.

5 References

1. Burnthal, E. V. et al.: Text. Chem. Color. 2 (13), 218 (1970)
2. Weidner, H. P.: Chemiefasern 18, 751 (1968)
3. Braun, P.: Chemiefasern 20, 39 (1970)
4. Albrecht, W.: Chemiefasern 18, 746 (1968)
5. Thimm, J. K.: Mell. 51, 177 (1970); JAATCC 2 (4) 69 (1970)
6. Hoffrichter, S. et al.: Deutsche Text. Techn. 20, 515, 585 (1970)
7. VEB Schwarza: Brit. 1.256.579 (1971)
8. Hoechst, Ger. 2.023.527 (1971)
9. Lacko, V.: Proc. Symp. Chemical Fibers, Praha (1965); Tabor (1966)
10. Kuraray Co.: Brit. 1.227.481 (1971); C.A. 75, 22396
11. Du Pont: U.S. 3.549.597 (1970)
12. Church, W. H., Shivers, J. C.: Text. Res. J. 29, 536 (1959)
13. Du Pont: U.S. 3.023.192 (1958)
14. Eastman Kodak: Fr. 1.359.090 (1964)
15. Leibnitz, E. et al.: Faserf. u. Textilt. 21, 426 (1970); 22, 541 (1971)
16. Leibnitz, E. et al.: Ger. (East) 73.115; C.A. 74, 32632
17. Asahi Chem. Ind.: Jap. 69 20469; C.A. 72, 13757
18. Baird, M. E., Hatfield, P., Morris, G. T.: J. Text. Inst. 47, T 181 (1956)
19. Driesch, H.: Mell. 37, 789, 921, 1034 (1956)
20. Gintis, D., Mead, E. J.: Text. Res. J. 29, 578 (1959)
21. Nüsslein, J.: Textilpraxis 16, 447 (1959)
22. Grünewald, K. H.: Chemiefasern 12, 853 (1962); 18, 862, 933 (1968)
23. Szegö, L.: Textilia 18 (4), 11 (1972)
24. Staudinger, H.: Mell. 18, 681 (1937); Ber. 70, 1565 (1937); 72, 1709 (1939)
25. Thimm, J. K.: Mell. 51, 177 (1970)
26. Berger, W. G.: Mell. 51, 438 (1970)
27. Montedison: Ger. 2.046.121 (1971); C.A. 75, 99213
28. Teijin: Jap. 70 36318; C.A. 74, 77328
29. Teijin: Jap. 70 32437; C.A. 74, 32629
30. Toray: Jap. 70 31701; C.A. 75, 7269
31. Mobil Oil Co.: U.S. 3.547.888 (1970); C.A. 74, 72882
32. Teijin: Jap. 71 05227; C.A. 76, 114195
33. Toray: Jap. 71 05225; C.A. 76, 114196
34. Jammers, H. C.: Mell. 50, 19 (1969)
35. Hall, D. M.: J. Appl. Pol. Sc. 15, 1539 (1971)
36. Teijin: Jap. 71 22170; C.A. 77, 71237
37. Toyo: Jap. 71 02769; C.A. 76, 4845

38. Kanegafuchi: Jap. 71 28981; C.A. *77*, 36228
39. Kanegafuchi: Jap. 70 30046; C.A. *75*, 7277
40. Allied Chem. Co.: U.S. 3.620.666 (1972); C.A. *77*, 21476
41. Teijin: Jap. 71 05558; C.A. *76*, 4827
42. Teijin: Jap. 70 34404; C.A. *75*, 37759
43. Hoechst: Brit. 1.173.392 (1969)
44. Kurashiki Rayon: Brit. 1.152.647 (1967)
45. Toyo: Jap. 71 33728; C.A. *77*, 63279
46. Teijin: Jap. 70 30448; C.A. *75*, 50337
47. Du Pont: U.S. 3.316.612 (1967)
48. Senner, P. et al.: Mell. *50*, 64 (1969)
49. McDowell, W. et al.: Mell. *50*, 59, 814, 1340 (1969)
50. Wecker, G.: Text. Praxis *27*, 117 (1972)
51. Dumbleton, J. H. et al.: Polymer *10*, 539 (1969)
52. Schroth, R. et al.: Faserf. u. Textilt. *22*, 273 (1971)
53. Coleman, D.: J. Pol. Sc. *14*, 15 (1954)
54. Griehl, W. et al.: Faserf. u. Textilt. *6*, 504, 554 (1955)
55. Kresse, P.: Faserf. u. Textilt. *11*, 353 (1960)
56. Teijin: Jap. 69 14952, 32310-2; C.A. *71*, 114080, *72*, 13271-3, C.A. *72*, 112691-4, 122806
57. Teijin: Jap. 70 04631, 07870; C.A. *73*, 36485, 46586
58. Teijin: Jap. 70 28595; C.A. *74*, 23573, *75*, 110966
59. Teijin: Jap. 71 13831, 18624; C.A. *76*, 114738, *77*, 21399
60. Rhodiaceta: Fr. 1.352.243 (1964); C.A. *61*, 767
61. Toyo: Jap. 68 21503; C.A. *71*, 62242
62. Goodyear: Ger. 2.011.050 (1970); C.A. *74*, 32651
63. Toray: Jap. 70 21596; C.A. *74*, 64458
64. Dow – BASF: Ger. 2.032. 818 (1971); C.A. *74*, 127370
65. Hoechst: Brit. 1.043.963 (1966); C.A. *66*, 11769
66. Teijin: Jap. 63 01981, 11496-7; C.A. *59*, 7702, *60*, 12165
67. Asahi: Brit. 1.076.877 (1967); C.A. *67*, 91619
68. Griehl, W.: Chemiefasern *16*, 775 (1966); C.A. *66*, 19652
69. Teijin: Jap. *65*, 542; C.A. *61*, 16086
70. Toyo: Jap. *67*, 8946; C.A. *68*, 30983
71. Eastman – Kodak: Fr.add. 86.225 (1966); C.A. *65*, 18745
72. Du Pont: Brit. 856.917 (1960); C.A. *55*, 18179
73. Mitsubishi: Jap. 67 16073; C.A. *68*, 22673
74. Teijin: Jap. 63 4798; C.A. *59*, 7702
75. Monsanto: Belg. 627.522 (1963); C.A. *60*, 12170
76. Griehl, W.: Chemiefasern *16*, 775 (1966); C.A. *66*, 19652
77. Edgar, O. B. et al.: J. chem. Soc. *1952*, 2633, 2638; C.A. *47*, 381-2
78. Edgar, O. B. et al.: J. Pol. Sc. *8*, 1 (1952); C.A. *47*, 9110
79. Reinisch, G. et al.: Eur. Pol. J. *6*, 205 (1970)
80. Jackson, J. B. et al.: Polymer *10*, 873 (1969); C.A. *71*, 125115
81. Jakob, F.: Chemiefasern *22*, 388 (1972)
82. Seda de Barcellona: Span. 286.873 (1963); C.A. *60*, 14690
83. Kuraray: Ger. 2.046.047 (1971); C.A. *75*, 50363
84. Snia Viscosa: Ger. 2.060.604 (1971); C.A. *76*, 89291
85. Du Pont: Fr. 1.546.544 (1968); C.A. *71*, 62237
86. Eastman – Kodak: U.S. 3.528.947 (1970); C.A. *73*, 110795
87. Toray: Jap. 70 32955; C.A. *75*, 37783
88. Teijin: Jap. 71 00847; C.A. *76*, 15676
89. BASF: Ger. 2.004.582 (1971); C.A. *76*, 35176
90. Hystron Fibers: U.S. 3.643.541 (1972); C.A. *76*, 128668
91. Union Carbide: U.S. 3.184.434 (1965); C.A. *63*, 4487
92. Goodyear: U.S. 3.554.975 (1971); C.A. *74*, 65475

93. Rhodiaceta: Ger. 1.912.582 (1969); C.A. *72*, 4293
94. Imp. Chem. Ind.: U.S. 3.226.360 (1965); C.A. *64*, 11395
95. Chem. Faser: Neth. 64 06003 (1964); C.A. *64*, 16076
96. Seda de Barcellona: Span. 338.761 (1968); C.A. *72*, 36766
97. Hoechst: Ger. 1.909.516 (1970); C.A. *74*, 4587
98. Siepmann, E.: Mell. *49*, 577 (1968)
99. Berger, W. G.: Mell. *51*, 438 (1970)
100. Beutler, H.: Mell. *51*, 1189 (1970)
101. Pfleger, J. et al.: Chemiefasern *22*, 1032 (1972)
102. Maerov, S. B. et al.: Text. Res. J. *31*, 697 (1961)
103. Rhodiaceta: Fr. 1.352.909 (1964); C.A. *61*, 8485
104. Imp. Chem. Ind.: Neth. 65 06421 (1965); C.A. *65*, 12336
105. Imp. Chem. Ind.: Neth. 65 09691 (1965); C.A. *64*, 19866
106. Teijin: Jap. 71 06432; C.A. *75*, 22368
107. Imp. Chem. Ind.: Ger. 2.031.743 (1971); C.A. *74*, 127362
108. Teijin: Jap. 70 38077; C.A. *74*, 88381
109. Toyo: Ger. 1.964.654 (1970); C.A. *73*, 67612
110. Teijin: Jap. 71 05226; C.A. *76*, 4855
111. Eastman – Kodak: Fr. 1.372.345 (1965); C.A. *62*, 13311
112. Imp. Chem. Ind.: Neth. 66 0812 (1966); C.A. *66*, 86550
113. Imp. Chem. Ind.: Fr. 1.484.108 (1967); C.A. *68*, 3843
114. Toray: Jap. 70 11831; C.A. *73*, 89081
115. Teijin: Jap. 70 19433; C.A. *74*, 4589
116. Teijin: Jap. 70 21267; C.A. *74*, 65450
117. Teijin: Jap. 70 34709; C.A. *75*, 37763
118. Teijin: Jap. 71 05222; C.A. *76*, 4858
119. U.S. Rubber: Neth. 65 0292 (1965); C.A. *64*, 8380
120. Toyo: Jap. 69 2027, 69 32303/6/7/14; C.A. *72*, 22549, 112693, 122829-30, 122844, 122855
121. Kurashiki: Jap. 70 11832; C.A. *73*, 89079
122. Medley, J. A.: J. Text. Inst. *45*, T 123 (1954)
123. Löbel, W.: Faserf. u. Textilt. *19*, 110 (1968); *23*, 385 (1972)
124. Valko, E. I., Tesoro, G. C.: Mod. Text. *38* (7), 62 (1957)
125. Method A Iso/TC – 38/SC, WG – 2.96 F or DIN 54 345
126. Schwenkhagen, H. F.: Mell. *34*, 1182 (1953)
127. Rösch, M.: Z. ges. Textilind. *63*, 968, 1054 (1901)
128. Reishaus, M.: Mell. *44*, 202 (1963)
129. Wegener, W. et al.: Text. Praxis *18*, 327, 417 (1963)
130. Quintilier, G. et al.: J. Text. Inst. *48*, P 26 (1957)
131. Hearle, J. W. S.: J. Text. Inst. *48*, P 40 (1957)
132. Mayer, F.: Textilveredlung *5*, 278 (1970)
133. Henry, P. S. H. et al.: J. Text. Inst. *58*, 55 (1967)
134. Romer, E. H. et al.: Text. Res. J. *38*, 28 (1968)
135. Wilkinson, P. R.: Mod. Text. *51* (12), 35 (1970)
136. Berg, H.: Mell. *52*, 448 (1971)
137. Reichle, A.: Mell. *50*, 1081 (1969)
138. Du Pont: U.S. 3.619.274 (1971); C.A. *76*, 73453
139. Toray: Jap. 71 26439; C.A. *77*, 21526
140. v. Hornuff, G.: Faserf. u. Textilt. *23*, 340 (1972)
141. Hoechst: Ger. 1.952.327 (1971); C.A. *75*, 37770
142. Du Pont: U.S. 3.619.274 (1971); C.A. *76*, 73643
143. Teijin: Jap. 71 21318; C.A. *77*, 7119
144. Kanegafuchi: Jap. 70 29913; C.A. *75*, 7362
145. Toyo: Fr. 1.526.402 (1968); C.A. *70*, 116170
146. Toyo: Brit. 1.179.206 (1970)

147. Toray: Ger. 2.006.810 (1970); C.A. *73*, 121456
148. Hercules Inc.: U.S. 3.515.698 (1970); C.A. *73*, 15775
149. Toray: Ger. 2.015.370 (1969); C.A. *74*, 32627
150. Toray: Brit. 1.214.484 (1970)
151. Jap. Synth. Chem. Ind.: Jap. 71 05223; C.A. *76*, 4828
152. Kanegafuchi: Jap. 71 07462; C.A. *76*, 47255
153. Toray: Jap. 71 07461; C.A. *76*, 60815
154. Teijin: Jap. 71 13829; C.A. *76*, 114719
155. Teijin: Jap. 71 13828; C.A. *76*, 114718
156. Toray: Jap. 71 15025; C.A. *76*, 128650
157. Rhodiaceta: Ger. 2.134.463 (1972); C.A. *76*, 128667
158. Asahi: Jap. 71 20260; C.A. *77*, 7116
159. Junichika: Jap. 71 18614; C.A. *77*, 21398
160. Kanegafuchi: Jap. 72 06750; C.A. *77*, 21440
161. Di Pietro, J. et al.: Text Res. J. *41*, 593 (1971)
162. Wandel, M. et al.: Chemiefasern *22*, 397, 527 (1972)
163. Finley, E. L. et al.: J. Fire Flammability *2* (10), 298 (1971)
164. Hendrix, J. E. et al.: J. Fire Flammability *3* (1) 2 (1972); C.A. *77*, 142199/200
165. Pfeifer, N.: Mell. *50*, 1229 (1969)
166. Tesoro, G. C.: Text. Res. J. *40*, 430 (1970)
167. Nametz, J. R. C.: Ind. Eng. Chem. *62* (3), 41 (1970)
168. Reeves, W. A., Guthrie, J. D.: U.S. 2.772.182 (1954)
169. Le Blanc, R. B.: Text. Chem. Color. *3*, 263 (1971)
170. Bilger, X. et al.: Teintex *29*, 837 (1964)
171. Stepnizka, H. et al.: J. Appl. Pol. Sc. *15*, 2149 (1971)
172. Staufer Chem. Co.: Ger. 2.131.040 (1972); C.A. *77*, 142340
173. Toray: Jap. 71 19756; C.A. *76*, 155481
174. Hoechst: Fr. 1.583.032 (1970); C.A. *73*, 36508
175. Linden, P.: Textilveredlung *6*, 651 (1971); C.A. *76*, 26313
176. Toray: Jap. 70 25989; C.A. *74*, 88470
177. Toray: Jap. 71 19183; C.A. *76*, 155465
178. Hercules Inc.: U.S. 3.645.962 (1972); C.A. *77*, 154831
179. Toray: Jap. 72 00621; C.A. *77*, 6047
180. Monsanto: U.S. 3.654.230 (1972); C.A. *77*, 20837
181. Allied Chem. Co.: Ger. 2.162.437 (1970); C.A. *77*, 166094
182. Toray: Jap. 71 37367; C.A. *77*, 63231
183. Akzona Inc.: U.S. 3.629.365 (1971); C.A. *77*, 128741
184. Celanese: Ger. 2.042.450 (1971); C.A. *77*, 128089
185. Fiber Ind.: Ger. 2.132.350 (1972); C.A. *77*, 128186
186. Eastman – Kodak: U.S. 3.624.033 (1971); C.A. *77*, 114213
187. Bowers, C. A., Chantry, G.: Text. Res. J. *39*, 1 (1969)
188. Peper, H. A., Berch, J.: Text. Res. J. *33*, 137 (1963); *34*, 29, 844 (1964), *35*, 252 (1965)
189. Peper, H. A., Berch, J.: Amer. Dyest. Rep. *54*, P 863 (1965)
190. Lewis, H. M.: Amer. Dyest. Rep. *57*, P 132 (1968)
191. Sontag, M. S.: Amer. Dyest. Rep. *58* (26), 19 (1969)
192. Shimauchi, Sh.: Amer. Dyest. Rep. *57* P 462 (1968)
193. Martin, C. L. et al.: Amer. Dyest. Rep. *60* (10) 38 (1971)
194. Kleber, R.: Text. Praxis *24*, 42 (1969)
195. Byrne, G. A.: J. Soc. Dyers Col. *88*, 66 (1972)
196. Kubelka, P., Munk, F.: Z. f. techn. Physik *12*, 593 (1931)
197. Huisman, M. A. et al.: Text. Res. J. *41*, 657 (1971)
198. Wlodarsky, G.: J. Text. Inst. *61*, 506 (1970)
199. Jacobasch, H. J.: Faserf. u. Textilt. *20*, 191 (1969)
200. Pinault, R. W.: Text. World *118* (5), 112 (1968)
201. Perry, E. M.: Amer. Dyest. Rep. *57*, P 405 (1968)
202. Monsanto: U.S. 3.410.927 (1968); C.A. *70*, 20956

203. Casella: Ger. 1.914.389 (1969); C.A. *72*, 56653
204. Deering-Milliken: Fr. 1.578.598 (1969); C.A. *72*, 112759
205. Moyse, J. A.: Textilveredlung *5*, 377 (1970)
206. Moyse, J. A.: Text. Inst. & Ind. *6*, 29 (1970)
207. Monsanto: U.S. 3.582.256 (1971); C.A. *75*, 50368
208. Imp. Chem. Ind.: Brit. 1.234.234 (1971); C.A. *75*, 62257
209. Atlas Chem. Ind.: Ger. 2.060.114 (1971); C.A. *75*, 99240
210. Toray: Jap. 71 06540; C.A. *76*, 26379
211. Beaunit: Ger. 2.062.547 (1971); C.A. *76*, 35308
212. Teijin: Jap. 71 22118; C.A. *77*, 7218
213. Liljemark, A. T. et al.: Text. Res. J. *41*, 732 (1971)
214. Reeves, W. A. et al.: Amer. Dyest. Rep. *57*, P 1053 (1968); *60* (4), 52 (1971)
215. Bobek, E.: Mell. *49*, 1213 (1968)
216. Du Pont: U.S. 3.236.685 (1966)
217. Amer. Cyanamid Co.: U.S. 3.284.364 (1966)
218. Deering – Milliken: U.S. 3.377.249 (1967); 3.535.141 (1970)
219. BASF: Ger. 1.912.180 (1970); C.A. *74*, 4598
220. Roehm & Haas: Ger. 2.007.925 (1970); C.A. *74*, 14165
221. Monsanto: U.S. 3.563.795 (1971); C.A. *74*, 113170
222. Smith, S. et al.: Text. Res. J. *39*, 441, 449 (1969)
223. Moreau, J. P. et al.: Amer. Dyest. Rep. *57*, P 683 (1968); *58* (3) 21 (1969)
224. Du Pont: U.S. 3.412.175 (1968); C.A. *70*, 20933
225. Pennsalt Chem. Co.: U.S. 3.427.332 (1969); C.A. *71*, 4467
226. Geigy: Fr. 2.009.407 (1970); C.A. *73*, 36506
227. Daikin Kogyo: Ger. 1.952.012 (1970); C.A. *73*, 57116
228. Hoechst: Ger. 1.900.234 (1970); C.A. *73*, 89104
229. Hoechst: Ger. 1.901.273 (1970); C.A. *73*, 110823
230. Stevens & Co.: U.S. 3.598.514/5 (1971); C.A. *75*, 141930/1
231. Monsanto: U.S. 3.582.257 (1971); C.A. *75*, 50370
232. Du Pont: U.S. 3.388.097 (1968); C.A. *69*, 28535
233. Allied Chem. Co.: Ger. 2.049.642 (1971); C.A. *75*, 37790
234. Feinauer, A. et al.: Amer. Dyest. Rep. *58* (11), 16 (1969)
235. Richardson, B. L.: Text. World *122* (10) 16 (1972)
236. Reichle, A.: Mell. *50*, 1081 (1969)
237. Meyer, F.: Textilveredlung *5*, 278 (1970)
238. Goodrich, H. F. et al.: Text. Chem. Color. *2*, 213 (1970); C.A. *73*, 56997
239. Viertel, O. et al.: Chemiefasern *20*, 216 (1970); C.A. *72*, 122782
240. Berg, H.: Mell. *52*, 448 (1971)
241. Wandel, M. et al.: Chemiefasern *22*, 397, 527 (1972)

Received June 9, 1978
H.-J. Cantow (editor)

A Theoretical Consideration of the Kinetics and Statistics of Reactions of Functional Groups of Macromolecules

N. A. Platé and O. V. Noah

M. V. Lomonosov State University of Moscow, Moscow, USSR

Table of Contents

Introduction

The study of the reactions of functional groups of macromolecules is one of the most rapidly growing domains in modern polymer science. There have been many important achievements in this field in last few years: on the one hand, experiments have yielded new data about the chemical reactions of functional groups of macro-molecules which allow the modification of natural and synthetic polymers giving them interesting new properties. On the other hand, the theory of macromolecular reactions has been further developed so that peculiarities in these reactions and the structure of the products obtained can be predicted.

In this paper the authors will try to survey the mathematical approaches involved in evaluating the kinetics of macromolecular reactions and the statistical properties of the products of these reactions. The results of experimental studies which have been reviewed in detail elsewhere [1-3] will not be discussed. Only reactions of func-tional groups of isolated macromolecules will be considered here.

Such reactions can be divided into two large groups:

A) reactions of functional groups with reagents of low molecular weight — "polymer-analogous reactions", and

B) reactions between functional groups of the same macromolecule — "intra-molecular reactions".

The consideration of both polymeranalogous and intramolecular reactions should take into account characteristic polymeric effects which are due to the chain struc-ture of macromolecules. The most important of these are [3-5]:

1. "The neighboring-groups effect", which is exhibited in the change of the mech-anism and kinetics of the reaction due to the change in the immediate environ-ment of the functional groups upon conversion.

2. Configurational effects leading to changes in the reactivity of functional groups due to the existence of a functional group of the same or different chemical or stereochemical structure at a neighboring site, and the consequent steric hindrances.

3. The alteration of the reaction rate depending on a change in the local concentra-tion of reactants in the vicinity of the macromolecule compared with the average volume concentration of the reactant.

4. Conformational effects which are due to a change in the chain's conformation in a given medium during the reaction.

5. Electrostatic effects due to the interaction of a charged macromolecule with particles of a reactant or of the charged functional groups with each other.

6. Supermolecular effects depending on the possible association or aggregation of species in solutions or in a solide phase.

Generally, the theory of macromolecular reactions should take into account all these effects. However, the consideration of even one of them is very difficult and becomes even more so when more than one of these effects is present.

Let us consider the possibilities of including some characteristic polymer effects in a theoretical description of macromolecular reactions.

A. Polymeranalogous Reactions

The simplest case of a polymeranalogous reaction is the reaction of functional groups with reactants of low molecular weight under conditions excluding the possibility of exhibiting any of the polymer effects listed above, i.e., a reaction in homogeneous conditions in a dilute solution with an excess of low molecular reactant and the polymer is stereoregular and conformational, electrostatic and neighboring-group effects can be neglected. We shall denote unreacted groups as A and reacted groups as B.

The kinetics of such a reaction obey the equation:

$$dP(A)/dt = -kP(A) \tag{1}$$

The solution of (1),

$$P(A) = P(A)_0 \exp(-kt) \tag{2}$$

gives the fraction of unreacted groups — $P(A)$ — at any time t (k is the constant of the reaction A → B).

The calculation of the distribution of lengths of sequences in the chain does not present any difficulties. At any intermediate moment, the product is a binary copolymer with a random distribution of A and B units. The probability of finding any sequence of units for such a copolymer can easily be calculated from the probabilities of finding A (or B) on a given site of the polymeric chain:

$$P(X_1, X_2, \dots X_n) = \prod_i P(X_i) \tag{3}$$

where $X_i = A$ or B with

$$P(A) + P(B) = 1 \tag{4}$$

The solution of kinetic Eq. (2) gives the complete description of the sequence length distribution.

For this simple case, the solution of the kinetic equation also gives an evaluation of the composition heterogeneity of the reaction products. Because of the stochastic and independent nature of the substitution processes A → B, the process can be regarded as a Markov chain of the zeroth order. From the general theory of regular Markov chains, it is known that the composition distribution is Bernoullian in this case, with dispersion defined as

$$\lim_{n \to \infty} D_n/n = \frac{(1 - P_{A/A}) P_{A/B} (1 - P_{A/B} + P_{A/A})}{(1 - P_{A/A} + P_{A/B})^3} \tag{5}$$

where n is the chain length and $P_{A/A}$ and $P_{A/B}$ are the Markov transitional probabilities of finding A on the right (or on the left) of A or B, respectively. For the Markov chain of the order 0, $P_{A/A} = P_{A/B} = P(A)$ and

$$\lim_{n \to \infty} D_n/n = P(A)P(B) \tag{6}$$

We now present the existing approaches to the calculation of the kinetics, sequence length distribution and composition heterogeneity for polymeranalogous reactions exhibiting some of the previously mentioned polymeric effects.

I Neighboring-Groups Effect

We consider the following reaction model: A units are transformed into B units; the reaction is irreversible and of the first order; and the reactivity of the A units depends on the nature of the nearest neighbors. We denote the rate constants of the transformation of A, having 0,1 and 2 reacted neighbors B, as k_0, k_1 and k_2, respectively. We must keep in mind the fact that these constants do not depend on the concentration of the reagents or on the degree of conversion.

1 Kinetics of the Process

In the mid-sixties a number of different approaches to the calculation of the kinetics of polymeranalogous reactions with neighboring-groups effects appeared in the literature[7-13]. The most exact solution was proposed by McQuarrie[13]. He defined two types of unreacted units sequences: a) the j-cluster: The sequence of j units A bordered by two units B and b) the j-tuplet: the sequence of j units A bordered either by A or by B units. If $P(BA_jB)$ is the probability of finding a j-cluster in a chain with N units and $P(A_j)$ is the probability of finding a j-tuplet in this chain, one can write:

$$P(A_1) = P(BAB) + 2\,P(BA_2B) + 3\,P(BA_3B) + \ldots + NP(BA_NB)$$
$$P(A_2) = P(BA_2B) + 2\,P(BA_3B) + 3\,P(BA_4B) + \ldots + (N-1)P(BA_NB)$$
$$\cdots\cdots\cdots\cdots \tag{7}$$
$$P(A_N) = P(BA_NB)$$

where N is the maximum length of the sequence of unreacted units; or, more concisely

$$P(A_j) = \sum_{i=0}^{N-j} (i + 1)P(BA_{j+i}B)$$

From Eq. (7) one can obtain inverse relationships:

$$P(BA_jB) = P(A_j) - 2\,P(A_{j+1}) + P(A_{j+2}) \tag{8}$$

Let us now consider the change of j-tuplet probabilities with time:

$dP(A_1)/dt = -k_2 P(BAB) - 2\ k_1 P(BA_2 B) - 2\ k_1 P(BA_3 B) - k_0 P(BA_3 B)$

$-\ 2\ k_0 P(BA_4 B) - 2\ k_1 P(BA_4 B) - \ldots = -k_0 P(A_3) - 2\ k_1 [P(A_2) - P(A_3)]$

$-\ k_2 [P(A_1) - 2\ P(A_2) + P(A_3)]$

$dP(A_2)/dt = -\ 2\ k_1 P(BA_2 B) - 2\ k_0 P(BA_3 B) - 2\ k_1 P(BA_3 B) - 2\ k_1 P(BA_4 B)$

$-\ 4\ k_0 P(BA_4 B) - \ldots = -2\ k_1 [P(A_2) - P(A_3)] - 2\ k_0 P(A_3)$

$dP(A_3)/dt = -2\ k_1 P(BA_3 B) - k_0 P(BA_3 B) - 2\ k_1 P(BA_4 B) - 4\ k_0 P(BA_4 B)$ (9)

$-\ 2\ k_1 P(BA_5 B) - 7\ k_0 P(BA_5 B) - \ldots = -2\ k_1 [P(A_3) - P(A_4)]$

$-\ k_0 [P(A_3) + 2\ P(A_4)]$

$\ldots\ldots\ldots\ldots\ldots$

$dP(A_j)/dt = -2\ k_1 [P(A_j) - P(A_{j+1})] - k_0 [P(A_j)(j-2) + 2\ P(A_{j+1})]$ $(j \geqslant 2)$

McQuarrie suggested trying to put these probabilities in the form:

$$P(A_j) = \exp(-jk_0 t)\ \psi(t) \quad (j \geqslant 2)$$ (10)

(Mitjushin[14]) proved that this expression is true).
It follows from (9) and (10) that

$$d\psi/dt = 2(k_0 - k_1)[1 - \exp(-k_0 t)]\psi(t)$$ (11)

The solution of Eq. (11) with initial condition $P(A_j)_{t=0} = 1$ (the initial product is a homopolymer) gives

$$\psi(t) = \exp\{2(k_0 - k_1)[t - (1 - e^{-k_0 t})/k_0]\}$$
$$P(A_j) = e^{-jk_0 t} \exp\{2(k_0 - k_1)[t - (1 - e^{-k_0 t})/k_0]\}$$ (12)

The expression (12) is a solution for all $j \geqslant 2$. To find $P(A_1)$, we use the equation

$$dP(A_1)/dt + k_2 P(A_1) = 2(k_2 - k_1)P(A_2) + (2\ k_1 - k_0 - k_2)P(A_3)$$ (13)

The solution of (13):

$$P(A_1) = e^{-k_2 t}\left\{ 2(k_2 - k_1)e^{\frac{2(k_1 - k_0)}{k_0}} \int e^{(k_2 - 2k_1)t} \cdot \right.$$

$$\cdot \exp\left[\frac{2(k_0 - k_1)}{k_0}\ e^{k_0 t} \right] dt + (2\ k_1 - k_0 - k_2)\ e^{\frac{2(k_1 - k_0)}{k_0}} \cdot$$ (14)

$$\left. \cdot \int e^{(k_2 - k_0 - 2k_1)t}\ \exp\left[\frac{2(k_0 - k_1)}{k_0}\ e^{-k_0 t} \right] dt + C \right\}$$

can be expressed by incomplete γ-functions:

$$\gamma(a;x) = \int_0^x u^{a-1} e^{-u} du$$

The final expression for $P(A_1)$ is:

$$P(A_1) = e^{-k_2 t} \left(1 - \frac{2(k_2 - k_1) \exp\{2(k_1 - k_0)/k_0\}}{k_0 [2(k_1/k_0 - 1)]^{(2k_1 - k_2)/k_0}} \left\{\gamma\left[\frac{2k_1 - k_2}{k_0}; 2\left(\frac{k_1 - k_0}{k_0}\right) e^{-k_0 t}\right]\right.\right.$$

$$\left.- \gamma\left[\frac{2k_1 - k_2}{k_0}; 2\left(\frac{k_1 - k_0}{k_0}\right)\right]\right\} - \frac{(2k_1 - k_0 - k_2) \exp\{2(k_1 - k_0)/k_0\}}{k_0 [2(k_1/k_0 - 1)]^{(2k_1 + k_0 - k_2)/k_0}} \tag{15}$$

$$\cdot \left\{\gamma\left[\frac{2k_1 + k_0 - k_2}{k_0}; 2\left(\frac{k_1 - k_0}{k_0}\right) e^{-k_0 t}\right] - \gamma\left[\frac{2k_1 + k_0 - k_2}{k_0}; 2\left(\frac{k_1 - k_0}{k_0}\right)\right]\right\}\right)$$

It should be pointed out that with a constant ratio $k_2 = 2k_1 - k_0$ (i.e., k_0, k_1, k_2 form an arithmetic progression) the equation for $dP(A_1)/dt$ changes to

$$dP(A_1)/dt = -2k_1 [P(A_1) - P(A_2)] - k_0 [2P(A_2) - P(A_1)]$$

i.e., it is reduced to the particular case of Eq. (9), and $P(A_1)$ in this case will also be defined by Eq. (12).

Thus we can determine $P(A_1)$ as a function of the time and the kinetic parameters of the process. Because $P(A_1)$ is the mole fraction of unreacted units of A in a chain, Eq. (12) is, in fact, the kinetic equation for polymeranalogous reactions.

Let us review other approaches to the solution of the kinetic problem[7-12]. The first successful attempt was made by Fuoss[7]. Although there is no final kinetic equation in his work, the suggested method of solution is the same as McQuarrie's[13] and thereby leads one to expect the same result.

The simplest and most convenient method of describing the kinetics of reactions exhibiting the neighboring-groups effect was suggested by Keller[8]. Denoting the average fractions of unreacted units with 0,1 and 2 reacted neighbors as N_0, $2N_1$ and N_2, respectively, the author derived for them the following equations:

$$dN_0/dt = -(k_0 + 2\bar{k})N_0$$
$$dN_1/dt = -(k_1 + \bar{k})N_1 + \bar{k}N_0$$
$$dN_2/dt = -k_2 N_2 + 2\bar{k}N_0 \tag{16}$$

where

$$\bar{k} = \frac{k_0 N_0 + k_1 N_1}{N_0 + N_1}$$

The solution of (16) with the initial conditions $(N_0)_{t=0} = 1$, $(N_1)_{t=0} = (N_2)_{t=0} = 0$ is:

$$N_0(\tau) = \exp\{-(2k + 1)\tau - 2(k - 1)(e^{-\tau} - 1)\}$$
$$N_1(\tau) = (e^\tau - 1) \exp\{-(2k + 1)\tau - 2(k - 1)(e^{-\tau} - 1)\}$$

$$N_2(\tau) = 2 e^{-k'\tau} \int_0^\tau e^{k'\tau}[1 - 2k + (k-1)e^{-\tau} + ke^\tau]N_0(\tau)d\tau \tag{17}$$

where $\tau = k_0 t$, $k = k_1/k_0$, $k' = k_2/k_0$.

Equations (16) are valid only when the initial product is a homopolymer[8]. The sum of expressions (17) is the mole fraction of unreacted units, i.e., it represents the solution of the kinetic equation.

The approach used by Alfrey and Lloyd[9] considers the number of sequences of i unreacted units bordered by reacted units N_i. The change of N_i with time can be described by the following equations:

$$dN_1/dt = -k_2 N_1 + 2 k_1 N_2 + 2 k_0 \sum_{n=3}^\infty N_n$$

$$dN_2/dt = -2 k_1 N_2 + 2 k_1 N_3 + 2 k_0 \sum_{n=4}^\infty N_n \tag{18}$$

$$\cdots\cdots\cdots\cdots$$

$$dN_m/dt = -2 k_1 N_m - (m-2)k_0 N_m + 2 k_1 N_{m+1} + 2 k_0 \sum_{n=m+2}^\infty N_n$$

Denoting the general number of units in a chain as S, we can write the fraction of unreacted units G as:

$$G = S^{-1} \sum_{n=1}^\infty n N_n$$

and the overall rate of the reaction is:

$$-SdG/dt = k_2 N_1 + 2 k_1 \sum_{n=2}^\infty N_n + k_0 \sum_{n=3}^\infty (n-2)N_n \tag{19}$$

The system (18) with Eq. (19) includes $(m+1)$ equations with $(m+2)$ unknown values — the N_n — and needs one more equation to be soluble. Alfrey and Lloyd[9] postulated that $N_{n+1} = 2 N_n - N_{n-1}$ supposing that this expression is more exact for larger n.

Arends[10] used a stochastic approach to the kinetic problem. He introduced the following parameters: β = the probability of finding B to the left of A; γ_u = the probability of finding B on the right of AA; and γ_r = the probability of finding B on the right of BA. Probabilities of different sequences can be expressed by these probabilities and the fraction of unreacted units f. For example:

$$P(BA_nB) = (1 - f)\beta\gamma_u(1 - \gamma_r)(1 - \gamma_u)^{n-2} \tag{20}$$

The change of probabilities of one, two or three units of A with time can be described either by equations of type (18) or by Eq. (20). This leads to the final kinetic equation:

$$dG/dt + k_2 G = -\exp\left\{\frac{-2(k_0 - k_1)}{k_0}\right\} \exp\left\{\frac{2(k_0 - k_1)}{k_0} e^{-k_0 t}\right\} \tag{21}$$

$$[2(k_1 - k_2) + (k_0 - 2k_1 - k_2)e^{-k_0 t}]$$

All three approaches considered above are based on various assumptions. The equations of Keller (16) are valid only when the fractions of A units converting into B with rate constants k_0 and k_1 are determined by ratios $N_0/(N_0 + N_1)$ and $N_1/(N_0 + N_1)$, respectively. One can show that this means that the probability of finding A or B to the right or to the left of the dyad AA does not depend on the units situated on the other side of this dyad. In fact, products of polymeranalogous reactions have such properties.

The accuracy of the solution of Alfrey and Lloyd depends on the accuracy of the relationship $N_{n+1} = 2N_n - N_{n-1}$.

The approach of Arends includes the assumption that the sequence length distribution is determined only by two independent parameters, namely, the probability of finding B to the left of A and the probability of finding B to the right of AA. In fact, it is true only for the distribution of unreacted units[14].

It is interesting to note that despite the absence of any rigorous grounds for such assumptions, the final kinetic equations for all these approaches can be expressed in the same form (as was shown by Keller[11]). One can also show that they are identical to the exact kinetic equation of McCarrie[3].

Lazare[12] proposed applying Bose-Einstein statistics to the calculation of the distribution of N_0 objects on N_1 cells (N_0 = the number of A units being in the centre of an AAA triad, N_1 = the number of A units present in the centre of AAB or BAA). Although the assumption about the random character of the distribution of sequences of unreacted units bordered by reacted ones is not self-evident, the final equations are identical to Keller's Eq. (16)[3, 12].

We discussed above only the simplest model, i.e., the irreversible reaction of the first order. Let us now consider how one can take into account some complications in the reaction. One such complication is to discuss non-first order reactions, for example, the chlorination of polyethylene which under certain conditions proceeds as a (1/2)-order reaction[15]. This resulting order can be explained by the following scheme[16]:

$$Cl_2 \xrightarrow{k^{(0)}} 2\,Cl\cdot$$

$$Cl\cdot + RH \xrightarrow{k^{(1)}} R\cdot + HCl$$

$$R\cdot + Cl_2 \xrightarrow{k^{(2)}} RCl + Cl\cdot \tag{22}$$

$$R\cdot + Cl_2 \xrightarrow{k^{(3)}} RCl$$

If one assumes that the effect of neighboring groups which is exhibited only at the propagation stage, is the same for both of these stages, i.e., $k_0^{(1)}/k_0^{(2)} = k_1^{(1)}/k_1^{(2)} = k_2^{(1)}/k_2^{(2)}$, then one can obtain the following equation:

$$-\frac{dRH}{dt} = Cl_2 \sqrt{\frac{k^{(0)}}{k^{(3)}}} \frac{k_0^{(1)} RH_0 + 2\,k_1^{(1)} RH_1 + k_2^{(1)} RH_2}{\left(\frac{k_0^{(1)}}{k_0^{(2)}} RH_0 + 2\,\frac{k_1^{(1)}}{k_1^{(2)}} RH_1 + \frac{k_2^{(1)}}{k_2^{(2)}} RH_2\right)^{1/2}} \qquad (23)$$

In fact, at low conversions when $RH \approx RH_0$, the reaction order is equal to $1/2$. Using Keller's approach, one can obtain a system similar to (16)

$$\frac{dN_0}{dt'} = \frac{-(k_0 + 2\,\bar{k})N_0}{(N_0 + 2\,N_1 + N_2)^{1/2}}; \quad \frac{dN_1}{dt'} = \frac{-(k_1 + \bar{k})N_1 + \bar{k}N_0}{(N_0 + 2\,N_1 + N_2)^{1/2}}$$

$$\frac{dN_2}{dt'} = \frac{-k_2 N_2 + 2\,\bar{k}N_1}{(N_0 + 2\,N_1 + N_2)^{-1/2}}$$

$$(24)$$

where

$$dt' = Cl_2 dt; \; k_0 = \left(\frac{k^{(0)}}{k^{(3)}} k_0^{(1)} k_0^{(2)}\right)^{1/2}$$

$$k_1 = \left(\frac{k^{(0)}}{k^{(3)}} k_1^{(1)} k_1^{(2)}\right)^{1/2}; \; k_2 = \left(\frac{k^{(0)}}{k^{(3)}} k_2^{(1)} k_2^{(2)}\right)^{1/2}$$

$$\bar{k} = \frac{k_0 N_0 + k_1 N_1}{N_0 + N_1}$$

Equations (24) were applied to describe the experimental data on the chlorination of polyethylene and some hydrocarbons of low molecular weight [3,16]. It was shown that experimental kinetics data are best described by curves which correspond to $k_0 : k_1 : k_2 = 1 : 0.38 : 0.11$ for polyethylene, and $1 : 0.35 : 0.08$ for n-hexadecane. These ratios are close to kinetic constants found from experimental data for the first-order reaction, chlorination of hydrocarbons[3].

Another possible complication of the model is to consider reversible reactions[17-23]. If we take into account the influence of the nature of the nearest neighbors on the reaction rate, then we must introduce six constants describing the possible mutual transformations of A and B units according to following scheme[17]:

$\sim A-A-A\sim \qquad \sim B-A-A\sim \qquad \sim B-A-B\sim$

$k_0 \downarrow\uparrow k_2' \qquad\quad k_1 \downarrow\uparrow k_1' \qquad\quad k_2 \downarrow\uparrow k_0'$

$\sim A-B-A\sim \qquad \sim B-B-A\sim \qquad \sim B-B-B\sim$

Following McCarrie's approach, one can write kinetic equations, for the probabilities of different sequences of units which are analogous to Eq. (9). In this case, as for the irreversible reaction, the right-handside of the equation for the j-tuplet probability will contain the probability of (j + 1)-tuplet. The resulting system of j equations with (j + 1) unknown parameters can be solved only with some additional relationship between the unknown parameters. For the irreversible reaction, such a relationship is[14]:

$$P(A_{j+1}) = P(A_j) \exp(-k_0 t) \qquad (j \geqslant 2)$$

Silberberg and Simha[17] assumed that for a reversible reaction

$$P(Y, X_{j-1}, B) = \frac{P(YX_{j-1})P(X_{j-1}B)}{P(X_{j-1})} \tag{25}$$

where X, Y can be either A or B.

For irreversible reactions when $X=A$, $j=3$. Silberberg and Simha assumed that $j=3$ for reversible reactions, too. Comparison of kinetic calculations for short chains performed both without any assumptions and with assumption (25) showed that this does not lead to significant errors[22].

The most important point in studying reversible processes is the analysis of the equilibrium state. For polymeranalogous reactions with the neighboring-groups effect, the equilibrium conditions can be written as

$$k_0 P(AAA) = k_2' P(ABA)$$
$$k_1 P(BAA) = k_1' P(BBA)$$
$$k_2 P(BAB) = k_0'(BBB)$$

Expressing the probabilities of various triads in terms of $P(A)$ and the transitional probabilities $P_{A \to B}$ and $P_{B \to A}$ where $P_{A \to A} = 1 - P_{A \to B}$; $P_{B \to B} = 1 - P_{B \to A}$, one can write[17]:

$$k_0 P(A) P_{A \to A}^2 = k_2' [1 - P(A)] P_{B \to A}^2$$
$$k_2 P(A) P_{A \to B}^2 = k_0' [1 - P(A)] P_{B \to B}^2$$
$$k_1 P(A) P_{A \to A} P_{A \to B} = k_1' [1 - P(A)] P_{B \to A} P_{B \to B} \tag{26}$$

It follows from this system of equations that

$$(k_0/k_2')(k_2/k_0') = (k_1/k_1')^2 \tag{27}$$

This equilibrium condition can also be derived from thermodynamic consider-ations. Vainstein, Berlin and Entelis[23] supposed that because of the energy change of neighbors which accompanied the change of a central unit of a triad, the analysis of a reversible reaction has to include ten equilibrium constants corresponding to ten different pathways. Among these ten constants, only four are independent, and the relation between them corresponding to the equilibrium state (27) can be derived by expressing all constants in terms of free energies and by taking into account the fact that some states are energetically equivalent.

One more possible complication of this model is to take into account the effect of more than two neighboring units. Krishnaswami and Vadav[27] considered the effects of four neighboring groups (two at the each side) and using the approach of Alfrey and Lloyd[9], obtained the kinetic equation with nine kinetic parameters.

2 Sequence Length Distribution

The model of the reaction under discussion can be regarded as a Markovian process with locally interacting components[14,25]. The process is Markovian in the sense that the state of the chain at the time $t + \Delta t$ depends stochastically on the state at the time t but not on the previous states. (This Markovian property in time should not be confused with the Markovian property in space, which is characteristic for the copolymerization process).

Some features of such processes can be used to describe the distribution of sequences of units. One of these features is the independence of finding sequences of A and B units on either side of the dyad AA. We shall not prove this theorem here, but we emphasize that from Eq. (12), it follows that for all $j \geqslant 2$ for every instant t:

$$P(A_{j+1})/P(A_n) = \exp(-k_0 t) \qquad (28)$$

In generalized form, $P(ZAAY)$ – the probability, of finding the sequence ZAAY (Z, Y represent any combination of the units A, B) – can be written as

$$P(ZAAY) = P(ZAA)P(AAY)/P(AA) \qquad (29)$$

It follows from this theorem that

$$P(ZA_{n+1}) = P(ZA_n) \exp(-k_0 t) \quad (n \geqslant 2) \qquad (30)$$
$$P(A_{i+1}ZA_j) = P(A_i ZA_{j+1}) = P(A_i ZA_j)\exp(-k_0 t) \quad (j, i \geqslant 2) \qquad (31)$$

The independence of finding any sequences to the right or to the left of the isolated unit A was shown elsewhere[14] in the particular case, when $k_2 = 2 k_1 - k_0$ (k_0, k_1 and k_2 form an arithmetic progression). In terms of probabilities, this property can be written as

$$P(ZAY) = P(ZA)P(AY)/P(A) \qquad (32)$$

These properties of the distribution of sequence in the products of macromolecular reactions are very important, and expressions (28)–(32) are widely used to describe them quantitatively.

The complete solution to the problem of describing the sequence lengths distribution includes the calculation of the probability of finding any sequence of n units A and B. Equation (12) allows one to calculate the probabilities of such sequences of unreacted units – n_A-tuplets. Evidently this is not sufficient for the complete description because of the significant difference in the distribution of reacted and unreacted units due to the irreversibility of the reaction.

Platé et al.[26-28] proposed a method of calculating the probabilities of all n_{AB}-tuplets (i.e., of sequences of both types of units), including the consideration of n_{AX}-tuplets constructed with the units A and X, where X represents a site which can be occupied either by A or by B.

The use of the idea of n_{AX}-tuplets is convenient because their probabilities can only decrease with time, and the probability of any sequence consisting of units A and B can be expressed with the aid of the probabilities of n_{AX}- and n_A-tuplets. For example:

$$P(ABAB) = P(ABA) - P(ABAA) = P(AXA) - P(AAA) - P(AXAA) + P(AAAA)$$

The simplest X-containing tuplet is AXA.

P(AXA) can decrease with time due to the fact that two fixed A units can be converted into B with rate constants k_0, k_1 and k_2 dependent on the nature of the neighbors. As the site X can be occupied either by an A unit or by a B unit which can be on either side of the run AXA, P(AXA) is composed of $2^3 = 8$ probabilities of different 5-tuplets:

$$P(AXA) = P(A_5) + P(A_4B) + P(BA_4) + P(BA_3B) + P(A_2BA_2) + P(BABA_2)$$
$$+ P(A_2BAB) + P(BABAB)$$

The resulting equation for dP(AXA)/dτ is:

$$dP(AXA)/d\tau = -2k'P(AXA) + 2(k'-k)P(AXA_2) + 2(k'-k)P(A_3)$$
$$- 2(k'-2k+1)P(A_4) \qquad (33)$$

where $\tau = k_0 t$, $k = k_1/k_0$, $k' = k_2/k_0$.

The same argument as led to (33) gives:

$$dP(AXA_2)/d\tau = -(2k+k')P(AXA_2) + (k-1)P(AXA_3) + (k'-k)P(A_2XA_2)$$
$$+ (k'-1)P(A_4) - (k'-2k+1)P(A_5) \qquad (34)$$
$$dP(A_2XA_2)d\tau = -4kP(A_2XA_2) + 2(k-1)[P(A_2XA_3) + P(A_5)] \qquad (35)$$

If we take expressions (30-31) into account, we obtain the closed system from Eqs. (33-35). The numerical solution of this system gives the probabilities of all sequences containing one B unit. The equations for probabilities of sequences with two, three and more X units can be derived in the same way.

The results of the calculation of probabilities of blocks of reacted units $P(AB_iA)$ for i = 1 - 3 are presented in Fig. 1[26-28]. It should be pointed out that these calculations are rather cumbersome and become more and more complicated with increasing sequence length. The only exception is the particular case of the constant ratio $k' = 2k - 1$ (i.e., when k_0, k_1 and k_2 form an arithmetic progression). In this case all the probabilities are determined analytically, and because of the

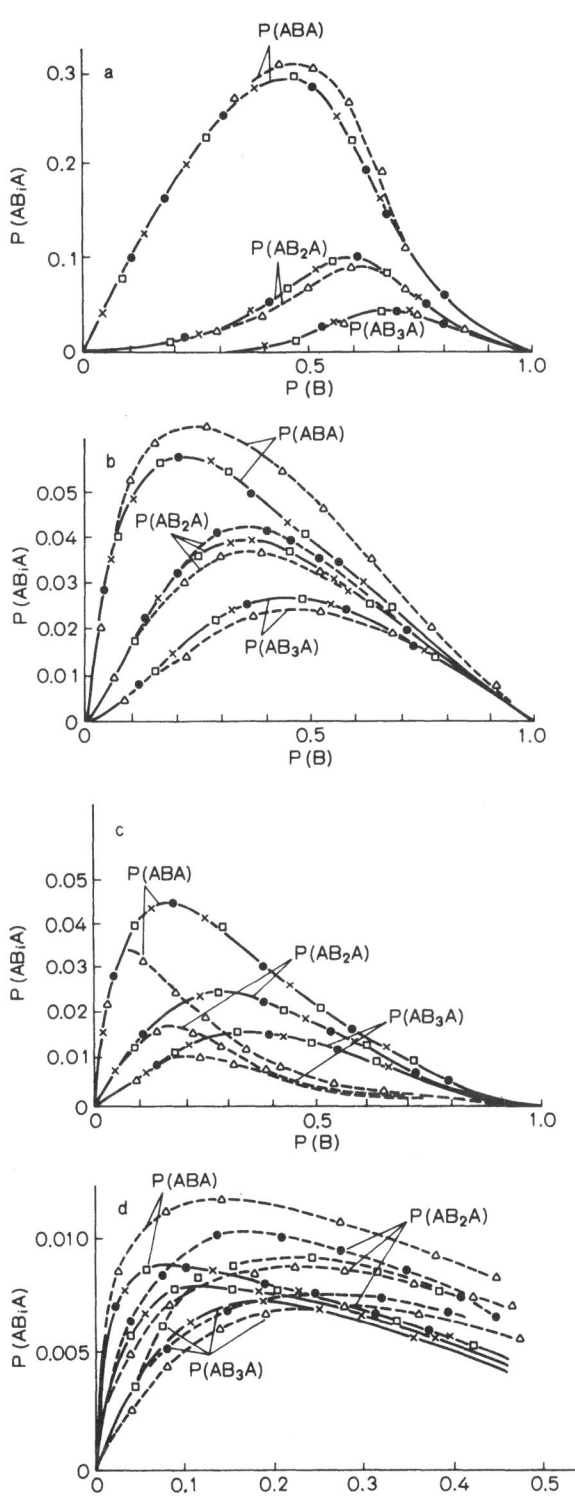

Fig. 1. Probabilities of blocks of i reacted units $P(AB_iA)$ vs. the degree of conversion $P(B)$ at $k_0:k_1:k_2 = 1:0.2:0.01$ (a), $1:5:5$(b), $1:5:100$ (c), $1:50:50$ (d): $(--)$ exact solution, (x) B-approximation ($\triangle$) first-order Markovian approximation, ($\bullet$) second-order Markovian approximation, ($\square$) third-order Markovian approximation[28]

validity of expression (32), the problem of the complete description of sequence distribution is simplified. In general, the exact calculation of all parameters of the distribution is so cumbersome that approximate methods of solving this problem are of special interest. One such approach is an approximation by Markovian chains of different orders[26-28].

To apply such an approximation, one supposes that at any fixed time the chain is n-th order Markovian, that is the state of any unit depends only on the state of n units to the left or to the right of it and does not depend on the state of the $(n + 1)$th unit. Moreover, one assumes that transformations in time between various states occur as polymeranalogous reactions with the neighboring-group effect.

In a first-order Markovian approximation, there are two independent conditional probabilities: $P_{A/A}$ (the probability of finding A to the right of A) and $P_{A/B}$ (the probability of finding A to the right of B), and all other characteristics of the chain structure are expressed in terms of them:

$$P_{A/A} = P(A_2)/P(A)$$
$$P_{A/B} = P(BA)/P(B) \tag{36}$$

Differentiation of (36) with respect to time results in

$$dP_{A/A}/d\tau = P_{A/A}[(k' - 2k) - P_{A/A}(k' - 2k + 1)(2 - P_{A/A})]$$

$$dP_{A/B}/d\tau = P_{A/B}\left\{ \frac{P_{A/A}^2}{1 - P_{A/A}} - k'(1 - P_{A/A}) \right.$$

$$\left. - P_{A/B}\left[\frac{P_{A/A}^2}{1 - P_{A/A}} + 2\,k\,P_{A/A} + k'(1 - P_{A/A}) \right] \right\} \tag{37}$$

The solution of the system (37) with initial conditions $P_{A/A} = P_{A/B} = 1$ at $\tau = 0$ gives a complete description of the chain structure in a first-order Markovian approximation.

In a second-order Markovian approximation there are four independent conditional probabilities: $P_{A/AA}, P_{A/BA}, P_{A/AB}, P_{A/BB}$; in a third-order, there are eight, and so on.

In Fig. 1 the results of Markovian approximations are compared with the results of exact analytical calculation. One can see that the use of Markovian approximations can be rather efficient in the case of a retarding neighboring-groups effect and in the case of small accelerations. The most optimal is evidently a second-order Markovian approximation, as the accuracy of a first-order approximation is not sufficient, while the accuracy of a third-order approximation is the same as the second order, but twice as many equations are required. The Markovian approximations are not very useful for greater accelerations, as they permit only the calculation of the probabilities of very short sequences with sufficient accuracy.

One more approximate approach to the calculation of units distribution in products of macromolecular reactions was proposed by Platé, Litmanovich and Noah[26-28]. This approach is based on the consideration of the kinetics of transformations of blocks of reacted units (the so-called B approximation).

Let us consider the block ABA. Such triads can arise from AAA triads with rate constant k_0, and disappear through transformation of the A units that are close to the unit B with rate constant k_1 (if this A unit is in the tetrad ABAA or AABA) and k_2 (if it is in the tetrad ABAB or BABA):

$$dP(ABA)/dt = k_0 P(A_3) - 2 k_1 P(ABA_2) + k_2 P(BABA) \tag{38}$$

The probabilities P(ABAA) and P(ABAB) can be expressed as

$$P(ABA_2) = P(ABA)P_{A/ABA}; \quad P(ABAB) = P(ABA)P_{B/ABA}$$

If one takes Klesper's assumption[29-30] about the independence of finding any sequences of units to the right or to the left of dyad BA

$$P_{Y/ZBA} = P_{Y/BA} \tag{39}$$

one can write

$$P_{A/ABA} = P_{A/BA}; \quad P_{B/ABA} = P_{B/BA} \text{ and so on,}$$

and Eq. (38) is transformed into

$$dP(ABA)/dt = k_0 P(A_3) - 2 \frac{P(ABA)}{P(AB)} [k_1 P(BA_2) + k_2 P(BAB)]$$

One can similarly derive the equation for $dP(AB_2A)/dt$ and so on.

For the blocks AB_jA, where $j \geqslant 3$ in the kinetic equation, one additional term arises describing the appearance of AB_jA due to the recombination of shorter blocks of the units B with rate constant k_2. The equations for the approximation can be written in a general form as follows:

$$\frac{dP(AB_jA)}{d\tau} = \delta_{j,1} P(A_3) + 2 k (1 - \delta_{j,1}) P(AB_{j-1}A) \frac{P(BA_2)}{P(BA)} \tag{40}$$

$$- 2 P(AB_jA) \frac{kP(A_2B) + k'P(BAB)}{P(AB)} + \frac{k'P(BAB)}{[P(BA)]^2} \sum_{m=1}^{j-2} P(AB_mA)P(AB_{j-m-1}A)$$

$$\text{where } \delta_{j,1} = \begin{cases} 1 \text{ when } j = 1 \\ 0 \text{ when } j \neq 1 \end{cases}$$

From the comparison of the results of B approximation calculations and the exact solution (Fig. 1), one can see that the results of this approximate method are close to the exact ones for a wide range of ratios of the constants. This method is also rather simple, and the numerical solution of Eq. (40) is not difficult, even when long sequences must taken into consideration.

So, accepting the assumption about the random distribution of blocks of reacted and unreacted units[29-30], one can calculate with the aid of Eqs. (40) and (12) the probability of any sequence of N blocks BA_jB and AB_iA as the product of probabilities of these blocks divided by $[P(AB)]^{N-1}$. For example:

$$P(AB_jA_nB_iA_mB) = \frac{P(AB_jA)P(BA_nB)P(AB_iA)P(BA_mB)}{[P(AB)]^3}$$

$P(AB_jA)$ are determined by Eqs. (40) and $P(BA_nB) = P(A_n) - 2P(A_{n+1}) + P(A_{n+2})$ by the expressions (12). The distribution of sequences of reacted and unreacted units corresponding to the model described above has thus been calculated.

The methods of calculation described above were used for the evaluation of triad composition of products of the hydrolysis of syndiotactic copolymers of methylmethacrylate and methacrylic acid in conditions corresponding to the accelerating[31] and retarding[32, 33] neighboring-groups effect. The probabilities of different triads calculated with constants being found from kinetic data ($k_0:k_1:k_2 = 1:3:5$[31] and $1:0,2:0,05$[32]) were compared with the results of NMR study[31−33]. The satisfactory coincidence of calculated and experimental values confirms the existence of the neighboring-groups effect in these reactions and allows the mathematical approaches we have developed to be applied to particular chemical reactions of polymers.

The next complication of the model to be considered is to take account of reversible reactions. The same thermodynamic approach used in deriving the equilibrium conditions[23] was proposed by Berlin, Vainstein and Entelis[34] for calculating the sequence distribution in polymeranalogous reaction products.

These authors assumed that the free energy of polymer chains is the sum of free energies of all units and of the energy which is due to the interchange of m reacted B units and n unreacted A units. Taking into account that each unit of the chain can have one of six different energy states with different values of the free energy,

AAA	BAA	BAB	ABA	BBA	BBB
F_1	F_2	F_3	F_4	F_5	F_6

one can obtain the following expression for F:

$$F = F_3 n_1 + F_4 m_1 + \sum_{i=2}^{\infty} [2F_2 + (i-2)F_1] n_i + \sum_{i=2}^{\infty} [2F_5 + (i-2)F_6] m_i$$

$$- RT \ln \frac{\left(\sum_{i=1}^{\infty} n_i\right)! + \left(\sum_{i=1}^{\infty} m_i\right)!}{\prod_{i=1}^{\infty}(n_i!) \prod_{i=1}^{\infty}(m_i)!} \tag{41}$$

where m_i, n_i are the concentrations of blocks of reacted and unreacted units of the length i.

After the minimization of F using the Lagrange method with the condition of material balance $\sum_{i=1}^{\infty} i(n_i + m_i) = n_0$ and the condition which is valid for infinite chains $\sum\limits_{i=1}^{\infty} n_i = \sum\limits_{i=1}^{\infty} m_i = u$, one can derive equations for sequence distribution in the chain:

$$n_1/n_0 = uba(\varphi_3\varphi_4)$$
$$n_i/n_0 = u(\varphi_3\varphi_4)(\varphi_2^2/\varphi_3)ba^i \qquad (42)$$
$$m_1/n_0 = uab^{-1}$$
$$m_i/n_0 = u(\varphi_5^2/\varphi_4)b^{-1}\varphi_6^{i-2}a^i$$

where $\varphi_i = \exp(-F_i/kT)$ and a and b are the Lagrange multipliers and n_0 is the initial concentration of monomeric units.

a, b and u are determined by following equations:

$$\left(\frac{\varphi_2^2/\varphi_3 a^2}{1-a} + a\right) \varphi_3\varphi_4 = \left(\frac{\varphi_5^2/\varphi_4 a^2}{1-a\varphi_6} + a\right)^{-1}$$

$$b = \left(\frac{\varphi_5^2/\varphi_4 a}{1-a\varphi_6} + 1\right) a \qquad (43)$$

$$\frac{n_0}{u} = 2 + \varphi_3\varphi_4\, a \left[\left(\frac{\varphi_5^2/\varphi_4 a}{1-a\varphi_6} + 1\right)\frac{\varphi_2^2/\varphi_3 a^2}{(1-a)^2} + \left(\frac{\varphi_2^2/\varphi_3}{1-a} a + 1\right)\frac{\varphi_5^2/\varphi_4 a}{1-a\varphi_6}\right]$$

As the φ_i are expressed through the equilibrium constants[23] the parameters of the distribution of sequences of units can be calculated in terms of these equilibrium constants or F values. On the other hand, the kinetic and thermodynamic parameters of the reaction can be found from the experimentally determined parameters of the distribution of units.

3. Composition Heterogeneity

If the calculation of sequence distribution in the products of polymeranalogous reactions is a more complicated problem than the kinetic description, the calculation of the composition heterogeneity is still more difficult. However, problems of this type can be solved using the Monte Carlo simulation method, i.e., with the aid of mathematical experiment.

In the publications of Platé, Litmanovich and Noah[35-38] Monte Carlo calculation of the composition heterogeneity was proved to be equivalent to the exact analytical solution, and its results could be used as a criterion of the accuracy of approximate analytical approaches. The polymeranalogous reaction can be simulated by computer. The conditions of the mathematical experiment (the optimum length of the model chain and the size of the sample population) were found from a comparison of calculated parameters of the distribution of units with exact values obtained by analytical solution. In Fig. 2 the results of such a comparison for $P(AB_iA)$ and "run number" $R = 2 P(AB)$ are presented. It can be seen from these figures that a chain length equal to 50 units and a number of chains equal to 50 are enough for the simulation retardation and slight acceleration effects. For larger accelerations ($k_0 : k_1 : k_2 = 1 : 5 : 100$) the lower

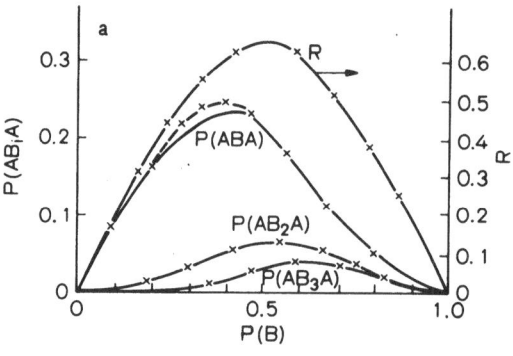

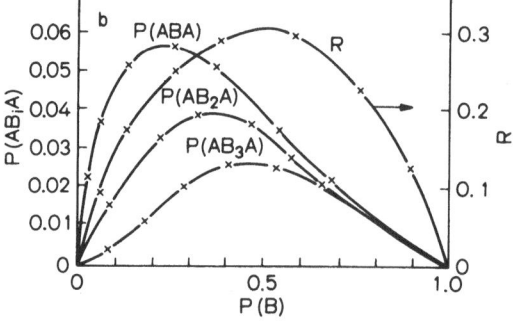

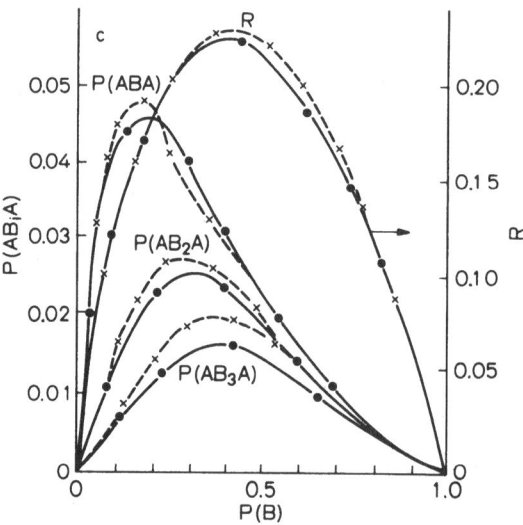

Fig. 2. Characteristics of the sequence length distribution versus the degree of conversion P(B): points, Monte Carlo calculation; curves, the accurate solution for $k_0 : k_1 : k_2 = 1 : 0.3 : 0.3$ (a), $1 : 5 : 5$ (b), $1 : 5 : 100$ (c); (×) Monte Carlo calculation for 100 chains with chain lengths of 50 units, (•) Monte Carlo calculation for 100 chains with chain lengths of 100 [38]

limit is a length of 100 units and 100 chains. The greater the accelerating effect the longer the minimum length of the model chain.

The results of the calculation of the parameters of composition heterogeneity, i.e., the functions of composition distribution and dispersion calculated as

$$D_n = \frac{n^2 \sum\limits_{i=1}^{\infty} (Y_i - \bar{Y})^2}{m} \tag{44}$$

where Y_i is the degree of conversion for the ith chain; $\bar{Y}$ is the average value of Y for all chains; n is the chain length; m is the number of chains, are presented in Figs. 3, 4, where they are compared with results of approximate calculations.

One approximate analytical approach to the calculation of composition heterogeneity was proposed by Frensdorf and Ekiner[39] for the products of the chlorination of polyethylene. These authors applied the Markovian approach to the description of the statistics of the substitution, assuming the independence of the probability of the reaction in an $(n + 1)^{th}$ unit on the state of $(n - 1)^{th}$ unit. This corresponds to a first-order Markovian approximation.

Introducing three independent parameters, the probability of substitution in the sequence of a unit more than one unit distant from other substituents (f), the probability of interaction with one nearest neighbor (μ), and that with two (ν), the authors expressed in terms of them the quantity $\phi(i, j, k)$, the conditional probability of finding j substituents in the nth unit, if in the $(n - 1)^{th}$ and $(n + 1)^{th}$ units, there are respectively i and k substituents ($i, j, k = 0$ or 1). Expressing $\phi (i, j, k)$ in terms of the Markovian transitional probabilities, Frensdorf and Ekiner obtained a system of equations which allow the calculation of the dispersion of composition distribution for polythene chlorination products.

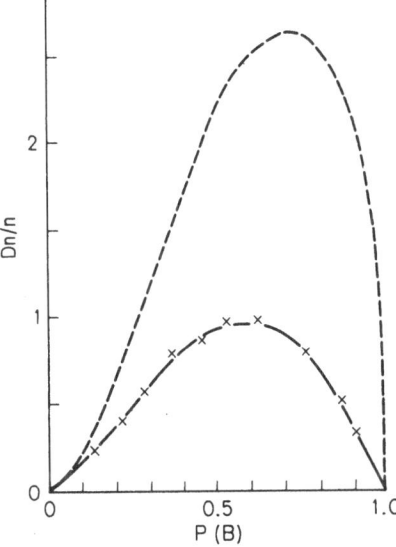

Fig. 3. Dispersion of composition distribution versus of conversion P(B) for $k_0 : k_1 : k_2 = 1 : 5 : 100$; broken curve, first-order Markovian approximation; solid curve, modified first-order Markovian approximation; points, Monte Carlo calculation[38]

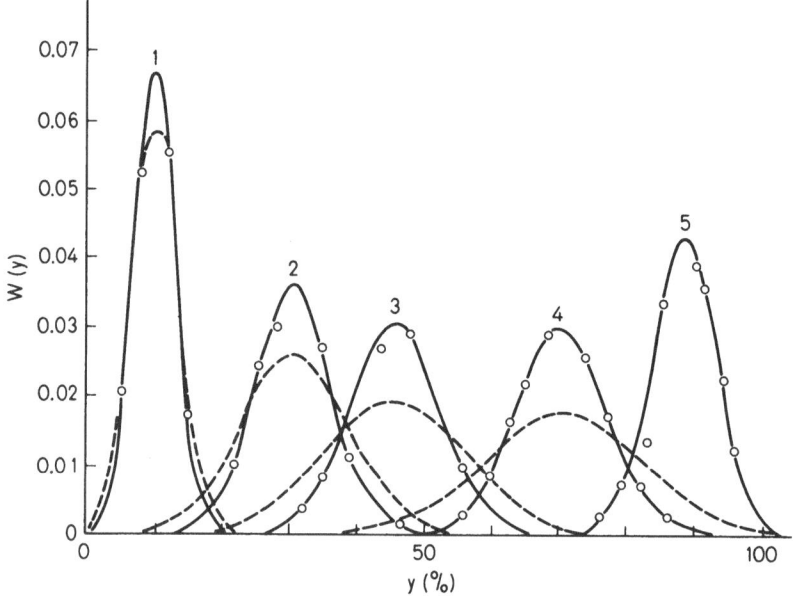

Fig. 4. Composition distribution functions $W(y)$ at $\bar{y} = 10\%$ (1), 30% (2), 46% (3), 71% (4), 89% (5) for $k_0 : k_1 : k_2 = 1 : 5 : 100$: broken curves, first-order Markovian approximation; solid curves, modified first-order Markovian approximation; points, Monte Carlo calculation[38]

After replacing the statistical parameters f, μ and ν by more convenient kinetic constants k_0, k_1 and k_2, one can use Eq. (37) for the calculation of the Markovian transitional probabilities.

It is known from the general theory of Markov chains[6] that the composition distribution for a first-order Markov chain approaches a Gaussian distribution, and that its dispersion is given by as:

$$\lim_{n \to \infty} \frac{D_n}{n} = \frac{(1 - P_{A/A})\, P_{A/B}\, (1 - P_{A/B} + P_{A/A})}{(1 - P_{A/A} + P_{A/B})^3} \tag{5}$$

$P_{A/A}$ and $P_{A/B}$, obtained from the solution of the system (37), allow the calculation of the function and the dispersion of composition distribution in the first-order Markovian approximation[37–38]. In Figs. 3, 4 the results of such calculations are compared with the results of Monte Carlo computation. It can be seen that in this case ($k_0 : k_1 : k_2 = 1 : 5 : 100$), the deviation is very strong, indicating the inaccuracy of this approximation.

Platé, Litmanovich and Noah[37, 38] proposed a modified form of a first-order Markovian approximation. In this approach, the notion of a Gaussian distribution form is maintained, and its dispersion can be calculated according to Eq. (5). But the Markovian transitional probabilities, $P_{A/A}$ and $P_{A/B}$, are proposed for calculating by the following expressions (36):

$$P_{A/A} = P(AA)/P(A); \quad P_{A/B} = P(BA)/P(B)$$

with $P(A)$, $P(B) = 1 - P(A)$, $P(AA)$ and $P(BA) = P(A) - P(AA)$ obtained from the solution of the exact Eqs. (12–14).

In other words, when a modified first-order Markovian approximation is formulated, one assumes that at the time under consideration, the chain is a first-order Markovian one, but all its previous history is described by the accurate equations.

In Figs. 3 and 4 the results of the calculation of D_n/n and of the composition distribution functions obtained by such a modified approximation are presented.

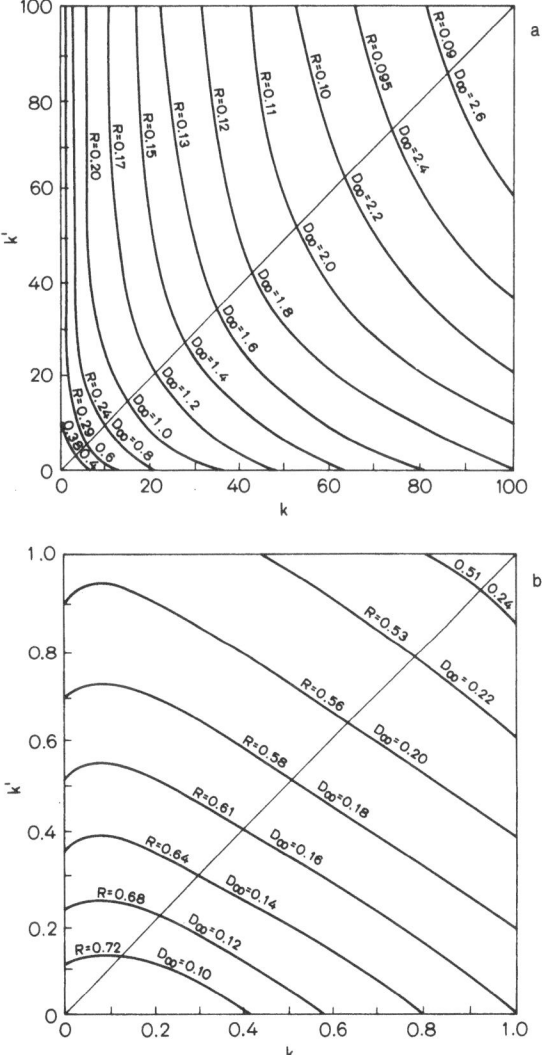

Fig. 5. $D_\infty = \lim\limits_{n \to \infty} (D_n/n)$ and $R = 2\,P(AB)$ dependence on $k_0 : k_1 : k_2$ for the accelerating effect, $k_1 \geqslant k_0$, $k_2 \geqslant k_0$ (a) and for the retarding effect, $k_1 \leqslant k_0$, $k_2 \leqslant k_0$ (b) at 50% conversion ($k = k_1/k_0$, $k' = k_2/k_0$)[38, 40]

The figures show that this method gives results very close to those of the mathematical experiment.

A modified first-order Markovian approximation is a rather simple method which allows the calculation of the dispersion of composition distribution for a wide range of ratios of the constants. The results of these calculations are summarized in Fig. 5.

In the framework of the approximation described above, one can find the interrelationship between the parameters of units distribution and composition heterogeneity. In fact, from expressions (5) and (36) one can obtain[3, 40]:

$$D_\infty = \frac{X(4X - R)}{R} \qquad\qquad R = \frac{4X^2}{D_\infty + X} \qquad\qquad (45)$$

where $X = P(A) \cdot P(B)$. That is, diagrams 5(a) and 5(b) also summarize the results of the calculation of the run number R.

One more approach to the calculation of composition heterogeneity for products of polymeranalogous reactions was proposed by Kuchanov and Brun[41]. These authors introduced for the description of the statistical properties of polymer chains of finite length N the N-dimensional stochastic vector $\mathbf{n}$, the components of which, n_j, are the values equal to the number of j-clusters in the chain. Introducing the distribution function of probabilities of the vector $\mathbf{n}$, $f(\mathbf{n}, t)$ and the generating function of this distribution $g(\mathbf{s}, t) = \sum_{\mathbf{n}} f(\mathbf{n}, t) \prod_{i=1}^{N} s_i^{n_i}$ (s_i are the components of auxiliary vector $\mathbf{s}$), and summing f over definite values $\mathbf{n}$, one can obtain some important characteristics of the chain structure. So the summation of f with the condition $\sum_i i n_i = m$ results in the function of the composition distribution of unreacted units, $f^*(m, t)$. The condition $\sum_i n_i = n$ results in the distribution functions of molecules on the total number of clusters n, $f(n, t)$. The summation on all values of components $\mathbf{n}$, except the jth, results in the function of distribution of the number of j-clusters.

The first moments of all these functions correspond to the average values of the composition and of the structure parameters. The second moments, determine the width of the corresponding distributions.

$$B_j(t) = \sum_{\mathbf{n}} n_j f(\mathbf{n}, t)$$

$$N_{i,j}(t) = \sum_{\mathbf{n}} n_i (n_j - \delta_{i,j}) f(\mathbf{n}, t) \qquad\qquad (46)$$

$$\delta_{i,j} = \begin{cases} 1 \text{ if } i = j \\ 0 \text{ if } i \neq j \end{cases}$$

With known moments (46) one can easily calculate other charactersitics of a chain structure. For a solution of Eq. (46) one can use the equation for the generating function $g(\mathbf{s}, t)$, which is derived from the densities of the probabilities of all clusters.

As $N \to \infty$, the authors obtained asymptotic expressions for composition heterogeneity, for example, which is described by the Gaussian law

$$\sqrt{\frac{N}{2\,\pi D_m}} \int_{\xi_1}^{\xi_2} \exp\left\{ -\frac{N(\xi - \overline{\xi})^2}{2\,D_m} \right\} d\xi = \frac{1}{2}\left\{ erf\left[\sqrt{\frac{N}{2\,D_m}}\,(\xi_2 - \overline{\xi}) \right] \right.$$

$$\left. -erf\left[\sqrt{\frac{N}{2\,D_m}}\,(\xi_1 - \xi) \right] \right\} \tag{47}$$

where ξ_1 and ξ_2 are the deviations from the average composition $\overline{\xi}$. $\overline{\xi}$ and D_m are calculated as the functions of f.

For some polymeranalogous reactions the results of the calculation of the composition heterogeneity could be compared with experimental data. So, for the quaternization reaction of poly-4-vinylpyridine with benzyl chloride, the constants ratio $(k_0 : k_1 : k_2 = 1 : 0.3 : 0.3)$ was calculated from kinetic data, and the functions of the composition heterogeneity calculated for this ratio were compared with data obtained by gel-permeation chromatography[42]. The comparison (Fig. 6) showed some additional effects (besides the neighboring-groups effect) which was demonstrated at high degree of conversion. In the same way (using the constants found from the kinetic data), the composition heterogeneity of the polyethylene chlorination reaction products was calculated and compared with the data of fractionation. The agreement of calculation with experiment confirmed the suggested mechanism for the reaction[43].

The mathematics for dealing with polymeranalogous reactions with the neighboring-groups effect is an area of the theory of macromolecular reactions in which a lot of work has been done and which could be applied to the study of particular chemical reactions of polymers. For some important reactions the kinetic constants could be found from experimental data[3, 31, 32, 42, 43] and used for the calculation of the parameters of sequence distribution[3, 31, 32] and composition heterogeneity[3, 42, 43]. The comparison of these parameters with corresponding experimental data led, in some

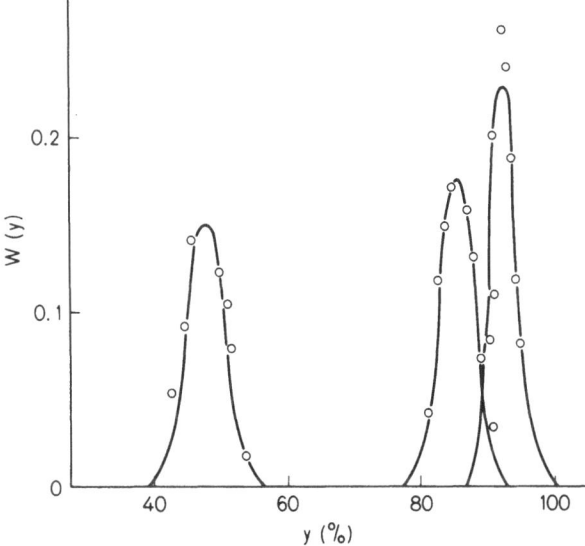

Fig. 6. Composition distribution for the reaction products of the quaternization of poly-4-vinylpyridine: points, experimental data; curves, calculation for $k_0 : k_1 : k_2 = 1 : 0.3 : 0.3$[38]

cases, to important conclusions about the reaction mechanism, and the possibility of the mathematical calculation of structure parameters not well-known from experiment allows the prediction of the properties of reaction products which depend on these parameters.

II Configurational Effect

The configurational effect is the dependence of the reactivity of a central unreacted A unit on the stereochemistry of the triad under consideration. There are three possible types of triads in vinyl polymers:

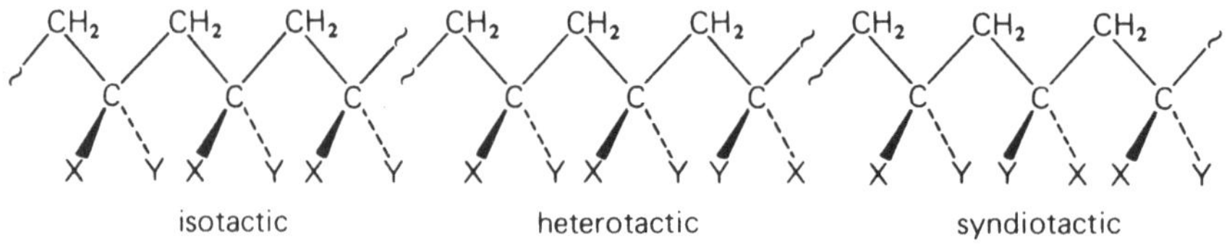

isotactic   heterotactic   syndiotactic

When the configurational effect is not accompanied by the neighboring-groups effect, the process can be considered to include three independent reactions with rate constants k^i, k^h and k^s. The calculation of the kinetics of such reactions is a rather simple task. But when the reactivity depends on both of these effects (which is more probable), we move from a problem with three parameters (k_0, k_1, k_2) to one with ten parameters, as an unreacted A unit can be in the centre of any one of the ten triads, and therefore can be converted into B with one of ten rate constants[3]:

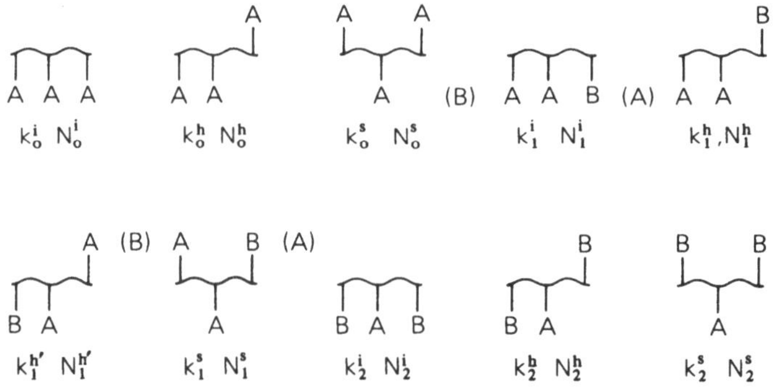

where N^i, N^h, N^s are the mole fractions of iso, hetero and syndio triads, and k^i, k^h, k^s are the rate constants of A units in the centre of the corresponding triads (the subscripts denote the number of reacted units).

In principle, this problem is very similar to the one considered above. So, for example, the kinetics of such a reaction can be described by the system of ten equations analogous to the Keller system (16)

$$-\frac{dN_0^i}{dt} = k_0^i N_0^i + 2 N_0^i \frac{k_0^i N_0^i + k_0^h N_0^h + k_1^i N_1^i + k_1^h N_1^h}{N_0^i + N_0^h + N_1^i + N_1^h}$$

$$-\frac{dN_0^h}{dt} = k_0^h N_0^h + N_0^h \left(\frac{k_0^i N_0^i + k_0^h N_0^h + k_1^i N_1^i + k_1^h N_1^h}{N_0^i + N_0^h + N_1^i + N_1^h} + \frac{k_0^h N_0^h + k_0^s N_0^s + k_1^{h'} N_1^{h'} + k_1^s N_1^s}{N_0^h + N_0^s + N_1^{h'} + N_1^s} \right)$$

$$(48)$$

and so on [3].

In the same way one can transform the equations describing the distribution of sequences and the composition heterogeneity.

While there is no single, outstanding complication in operating with ten constants, it is rather difficult to apply Eq. (40) to particular chemical reactions because of the experimental difficulties in evaluating the rate constants. This problem is rather complicated even for the stereoregular samples when one needs to determine only three constants [3]. The application of the approach worked out for stereoregular polymers to atactic ones includes the synthesis of iso- and syndiotactic models of high regularity.

It would appear that the calculation of the kinetics and the statistics of reactions in atactic chains could be simplified if some of the ten constants were equivalent or if there were some other restrictions. Pismen [44] calculated the kinetics of a polymer-analogous reaction in atactic chains for the particular case: $k_0^i = k_0^h = k_0^s = k_0$, $k_1^h = k_1^s = k_1$, $k_1^i = k_1^{h'} = k_2$, $k_2^i = 2 k_2 - k_0$, $k_2^h = k_1 + k_2 - k_0$, $k_2^s = 2 k_1 - k_0$ and $k_2 > k_1 > k_0$ (autoaccelerating reaction).

In this case one can see three types of j-clusters (i.e. sequences of j unreacted units bordered by reacted units): open, semiopen and closed, depending on the number of ultimate unreacted units in iso-position to the reacted neighbor (2-open, 1-semiopen, 0-closed, respectively).

The consideration of the rate of change with time of these three types of clusters results in a system of kinetic equations, the solution of which at $k_2 \gg k_1, k_0$ gives the fraction of unreacted units $P(A)$:

$$P(A) = \left(\frac{1 - \chi \exp(-k_0 t)}{1 - \chi} \right)^{\frac{2}{\chi} \left[\frac{k_1}{k_0} (1-\chi) - 1 \right]} \exp[-(2 k_i - k_0)t]$$

As $\chi \to 0$ (stereoregular polymer) this expression coincides with expression (12) for $j = 1$ (it is true for $k_2 = 2 k_1 - k_0$). Therefore one can also calculate the kinetics of polymeranalogous reactions in atactic chains.

The quantitative account of other macromolecular effects during polymeranalogous reactions is as yet an unsolved problem which should be the object of further research.

B. Intramolecular Reactions

Intramolecular reactions, i.e., reactions of functional groups of one macromolecule with or without agents of low molecular weight can be divided into three types.

The first type includes the reaction which takes place predominantly between adjacent groups. Those are, for example, the stripping of chlorine by Zn from poly(vinyl chloride), aldole condensation of poly(methyl vinyl ketone) and so on[45–58]. The kinetics and statistics of such reactions are described by almost the same mathematics as polymeranalogous reactions. The second type includes the reactions of intramolecular catalysis. These reactions depend on the conformation of a macromolecule because reacting groups can be isolated from each other along the chain, and the probability of their interaction is connected with the flexibility of the polymer chain. Another peculiarity of these reactions is the relative stability of conformation of the macromolecule during the reaction. The third type is the reaction of irreversible intramolecular cross-linking – a process which is also determined by chain conformation, but, in addition, the formation of each cross-link results in the change of the conformation of the macromolecular coil.

I Reactions between Adjacent Groups

The first theoretical treatment of these reactions was due to Flory who in 1939 calculated the average number of groups which remained unreacted until the end of the reaction[45].

Denoting the average number of unreacted groups in the chain run of n units as S_n, Flory wrote the following obvious relations:

$$S_0 = 0, S_1 = 1, S_2 = 0, S_3 = 1;$$

$$S_4 = \frac{2 S_2 + 2 S_1}{3}; \quad S_5 = \frac{2 S_3 + 2 S_2 + 2 S_1}{4}; \quad \ldots\ldots$$

$$S_n = \frac{2}{n-1} (S_1 + S_2 + \ldots + S_{n-2}).$$

Introducing the new variables $\Delta_n = S_n - S_{n-1}$, one can write

$$\Delta_n = 1 - \frac{2}{1!} + \frac{4}{2!} - \frac{8}{3!} + \ldots \frac{(-2)^{n-1}}{(n-1)!}$$

As $n \to \infty$, $\Delta_n \to \Delta_\infty = 1/e^2$ and $S_n \approx n/e^2$. That is, the fraction of isolated unreacted groups in rather long chains is roughly equal to 0.1353.

After Flory other authors also looked into the kinetics of intramolecular reactions[46–58]. Cohen and Reiss[46] used the method of multiplets, which was mentioned above in connection with polymeranalogous reactions[13]. Introducing the number of n-tuplets in j^{th} chain at the instant t, $C_n^{(j)}(t)$ (recall that an n-tuplet is a sequence of n unreacted units bordered by reacted or unreacted units), one can write:

$$-dC_n^{(j)}/dt = k[(n-1)C_n^{(j)} + 2\, C_{n+1}^{(j)}] \tag{49}$$

where k is a constant of the intramolecular reaction.

The solution of Eq. (49) as $n \to \infty$ leads to the following expression for the probability of n-tuplet:

$$P_n(t) = \exp[-(n-1)kt]\exp[-2(1-e^{-kt})] \tag{50}$$

For n = 1,

$$P_1(t) = \exp[-2(1-e^{-kt})] \tag{51}$$

Equation (51) is the final kinetic equation for the intramolecular reaction, while Eq. (50) describes the sequence length distribution of unreacted units. As $t \to \infty$, $P_1 \to 1/e^2$ corresponding to the classical result of Flory.

McQuisten and Lichtman[49] used another approach to the solution of this kinetic problem. They supposed that paired interactions in the polymer chain can be simulated by the process of random throwing of a "dumb-bell" consisting of two units on the one-dimensional lattice of N cells. After m trials with the density of throwing on one unit per unit time, $\nu = m/Nt$, the degree of filling θ, is equal to

$$\theta(t) = 1 - \exp[-2(1-e^{-\nu t})]$$

at $\nu = k$ this expression is identical to Eq. (51).

The same result was obtained by Barron and Boucher[51], using an approach analogous to that of Alfrey and Lloyd[9].

If N_x is the number of sequences of x unreacted units bordered by reacted ones, m the number of units in the chain, M_m the total number of chains, then one can write the following system for N_x:

$$dN_x/dt = 2k \sum_{i=x+2}^{m} N_i - k(x-1)N_x \quad (x \leqslant m-2)$$

The solution of this system gives an expression for the degree of conversion which coincides with Eq. (51).

Boucher applied the same approach to the calculation of the kinetics of intramolecular reactions with neighboring-group effects[52]. Introducing three rate constants k_0, k_1, k_2, as in the case of polymeranalogous reactions, he obtained the following equations:

(with $\tau = k_0 t$, $k = k_1/k_0$, $k' = k_2/k_0$)

$$\lim_{m \to \infty} \frac{N_x}{mM} = \begin{cases} 2 \int_{\exp(-\tau)}^{1} \{[k(1-u)^2 u^{2k-1} + (1-u)u^{2k}]\exp[(k-3)+2u \\ \quad + (1-k)u^2]\} \, du \quad (x = 1) \\[2mm] 2 e^{-k'\tau} \int_{\exp(-\tau)}^{1} \{[k(1-u)^2 u^{2k-k'} + (1-u)u^{2k-k'-1}] \\ \quad \exp[(k-3)+2u+(1-k)u^2]\} \, du \quad (x = 2) \\[2mm] (1-e^{-\tau})^2 \exp[-\tau(x-2k-3)]\exp[(k-3)+2e^{-\tau}+(1-k)e^{-2\tau}](x \geqslant 3) \end{cases}$$

The total fraction of unreacted units is

$$P_1 = \lim_{m \to \infty} \sum_{i=1}^{m} iN_i/mM = \lim_{m \to \infty} \left(\frac{N_1 + N_2}{mM}\right) + (3 - 2 e^{-\tau})$$

$$e^{-2k\tau} \cdot \exp\left[(k-3) + 2 e^{-\tau} + (1-k)e^{-2\tau}\right]$$

Boucher also showed that the same analysis can be used when the reaction proceeds between three[53] or n[54] functional groups.

Gonzalez and Hemmer[58] considered the particular case where the reaction is interrupted at some degree of conversion. Then the bonds which block some part of the unreacted groups are destroyed and the reaction proceeds.

II Intramolecular Catalysis

Intramolecular catalysis reactions were studied theoretically by Morawetz et al.[59-63] and by Sisido[64-68].

The simplest case of reactions between functional groups which are situated apart one from another are the reactions of cyclization, i.e., the reactions of end groups of one macromolecule. The interaction between such groups is possible only if they are close to one another. The probability of such proximity can be calculated using the end-to-end distribution function.

For the Gaussian model of the polymer chain of Z segments of length b, the probability of the distance between the ends being equal to h is determined by the function[60, 61].

$$W(h)dh = (2 \pi Zb^2/3)^{-3/2} \exp\left[-3h^2/2 \pi Zb^2\right] 4 \pi h^2 dh \tag{52}$$

$W(h)/4 \pi h^2$ denotes the concentration and tends to some limit at $h \ll Zb^2$. This limiting value $\lim[W(h)/4 \pi h^2] = C_{ef}^\circ$ can be regarded as the average concentration of one end of the polymer chain in the vicinity of the other end. In mol/l units, C_{ef}° is expressed as

$$C_{ef}^\circ = (1\,000/N_A)(3/2 \pi \bar{h}^2)^{3/2} \tag{53}$$

where $\bar{h}^2$ is the mean square distance between the ends and N_A is Avogadro's number.

If k_2 is the rate constant of the second-order reaction of functional groups of the polymer not fixed to the chain, the rate constant of the first-order reaction between the end groups of the chain is

$$k_1 = k_2 C_{ef}^\circ = k_2(1\,000/N_A)(3/2 \pi \bar{h}^2)^{3/2} \tag{54}$$

When the effect of the excluded volume can be neglected, the probability of the reaction between the group being on the n-th site of the polymer chain and catalytic group being on the j-th site can be estimated in the same way as for the reaction of end groups of the chain of $(j - n)$ segments. Then the mean square distance between the n-th and j-th units can be expressed through mean square end-to-end distance as

$$\overline{h}_{nj}^2 = \overline{h}^2 |j - n|/z \text{ for } |j - n| \geqslant x'$$

where x' is the minimum number of monomer units allowing the interaction.

The effective local concentration of the catalyst near the n-th reactive segment of the chain of Z segments with random distribution of catalytic groups can be written as

$$C_{ef} = (1000/N_A)(3 Z/2 \pi \overline{h}^2)^{3/2} \left(\sum_{x=1-n}^{x'} |P_j x|^{-3/2} + \sum_{x=x'}^{Z-n} |P_j x|^{-3/2} \right) + C'_{ef} \quad (55)$$

where $x = j - n$ and $P_j = 1$ or 0, depending on the presence of catalytic groups on the jth site: C'_{ef} is a correction parameter which accounts for the probability of the interaction of groups which are divided into less than j monomer units.

The rate constant is given by $k_1 = k_2 C_{ef}$. If the number of reactive groups is small and all other segments have catalytically active substituents, $P_j = 1$ for all j in Eq. (55) and $k_1 = k_2 C_{ef}^{max}$. If the fraction of catalytic groups is equal to W, $k_1 = W k_2 C_{ef}^{max}$.

After integration of Eq. (55) and substitution of $\overline{h}^2 = Z(K_\Theta/\phi)^{2/3} M_0/2$ (where $K_\Theta = [\eta]_\Theta \cdot M^{-1/2}$, $\phi = 2.6 \times 10^{21}$ is the Flory constant, and M_0 is the molecular weight of a monomer unit), one obtains[61]:

$$k_1 = W k_2 (4000 \phi/N_A K_\Theta)(3/2 \pi M_0)^{3/2} (x')^{-1/2} \quad (56)$$

All these calculations correspond to θ conditions, while taking account of the excluded volume results in the decrease of the probability of cyclic conformation. In that case $k_1 \sim Z^{-a}$, where $a > 3/2$.

The experimental study of the hydrolysis of the ternary copolymer of acrylamide with small amounts of reactive monomers I or II and catalytically active III showed that there is an intramolecular catalysis in such a reaction, but that the rate constant decreases with the decrease in the thermodynamic quality of the solvent[61].

Goodman and Morawetz[60] also used a computer to simulate the kinetics of intramolecular reactions for chains with some fraction W of catalytically active substituents. The population of 100 chains of 1000 units was considered with the assumption that there is only one reactive group in the middle of the chain. Only the interaction between groups separated by more than ten units was allowed.

The rate constant was calculated as follows:

$$k_n = C \left[\sum_{j=1}^{490} P_{nj}(500 - j)^{-a} + \sum_{j=510}^{1000} P_{nj}(j - 500)^{-a} \right] \tag{57}$$

where $P_{nj} = 1$ if there is a reactive group on the jth site and $P_{nj} = 0$ if there is no reactive group on the jth site.

The fraction of reactive groups was calculated as

$$Y(t) = \frac{1}{1000} \sum_n \exp(-k_n t) \tag{58}$$

The experimental dependences of log y on t were compared with curves calculated at various a. It was shown that the kinetics of the reaction in a good solvent was described by the expression with a = 2, but close to the θ conditions a is equal to 1.6 (this value is more than that for the model with zero excluded volume).

Sisido[64−68] studied the influence of catalytically active groups distribution in the chain on the rate of the reaction. He considered three cases:
1) The catalytic groups distribution is averaged for all molecules and can be neglected in the kinetic calculation.
2) The catalytic groups are fixed in the polymer chain and the probability of finding a catalytic group on a given site, independent on the state of neighboring groups is given.
3) The probability of finding a catalytic group on a given site depends on the state of neighboring groups.

All calculations were performed under the assumption that only groups separated by Z or more units can react. The equations were derived for the calculation of the fraction of unreacted units, depending on the catalytic groups distribution and on the constant of the interaction of groups separated by i units $-k_i$. The results of the calculation were compared with experimental data of Goodman and Morawetz[60]. The agreement of the experimental data with the curves calculated for a = 2 was demonstrated.

The probability of the cyclization for short chains was also calculated by the Monte Carlo method[65]. The results of these calculations agreed satisfactorily with experimental data[64, 65].

III Intramolecular Cross-Linking

The reactions of intramolecular cross-linking is an area ill suited to theoretical analysis. That is why despite the interest both from the view-point of the theory of macromolecular reactions and the theory of network formation there are very few papers concerning the theoretical study of this problem.

1 Dimensions of Cross-Linked Macromolecules

The first paper dealing with the calculation of dimensions of macromolecular coils containing cross-links is probably that of Zimm and Stockmayer. In 1949 these authors calculated the mean-square radius of gyration for the model of a free-rotational cyclic chain[69]. It was shown that

$$\overline{R}_{cyc}^2 = Nb^2/12 \qquad (59)$$

while for the linear chain

$$\overline{R}_{lin}^2 = Nb^2/6 \qquad (60)$$

is valid. (N is the number of segments in the chain; b is the length of a segment). Then they assumed that the contribution of the cyclic and linear parts to the value of the overall radius of gyration is proportional to the respective number of units in them, i.e.,

$$\overline{R}^2 = (Z/N)(Zb^2/12) + [(N-Z)/N](N-Z)b^2/6 \qquad (61)$$

where Z is the number of units in the cycle.

The approach of Zimm and Stockmayer allows the calculation of the dimensions of macromolecules having one cycle, i.e., with one cross-linkage. It is impossible to apply this approach to the chains with large numbers of cross-links because of the rapid increase of the number of possible topological structures accompanied by the complication of the algorithm of the calculation of $\overline{R}^2$ for each of them. There are three possible structures for the chain with two cross-linkages; for three cross-linkages, there are eight structures, and so on[70]:

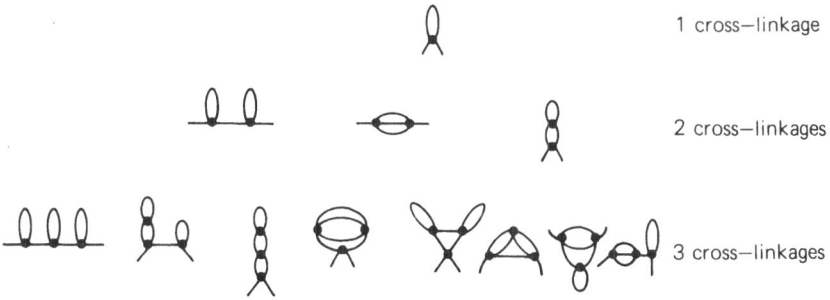

Edwards et al.[71, 72] used a thermodynamic approach to the problem of the calculation of the dimensions of cross-linked macromolecules. In[71] the general thermodynamic theory of polymer chains with cross-linkages was published. The equation of state of the cross-linked polymer is derived using the method of second-ary quantization, allowing operation with the average number of cross-links in the chain, neglecting the details of the chain structure. The theory[72] is applied to the calculation of dimensions of chains with intramolecular cross-linkages. The model considers isolated chains with sites of cross-linking distributed at random. It is further assumed that the chain is a free-rotational one, and the volume effects and van der Waals interactions between units are taken into account. The mean-square end-to-end distance is calculated as $\overline{h}^2 = C_N Nl^2$, where N is the chain length, l is the length of the unit, and C_N is the so-called characteristic ratio. The volume interactions are taken into account by introducing the "thickness" of the chain a (the square root of the area of a cross section of the chain). The cross-linkage is considered to be an equilibrium form with limited dimensions (two cross-linked units are a distance b_2 apart).

The calculation of the thermodynamic potential of a cross-linked chain, assuming the energetic advantage of cross-linking, leads to the following relationship between chain dimensions and the number of cross-linkages m:

$$\frac{1 + m\zeta^3/Nla^2}{1 - (1 + m\zeta^3/Nla^2)\tau^{-3}} = \frac{V}{kT} \frac{1}{2la^2} \left[1 - \frac{10\,\pi\,b_1^2}{3(Nla^2)^{2/3}\,\tau^2}\right] - \frac{\pi C_N l^2 \tau}{3(Nla^2)^{2/3}}$$

$$+ \frac{m+1}{N}\tau^3\left[1 - \frac{4\,\pi\,b_2^2}{3(Nla^2)^{2/3}\tau^2}\right] \qquad (62)$$

where τ is an intrinsic volume defined by $\tau^3 = (2\,R/3)^3(Nla^2)^{-1}$, ζ^3 is the volume of one cross-linkage, b_1 is the distance between units interacting with the attracting force determined according to the Flory-Huggins theory, and V is a change of inter-action energy due to the formation of the additional polymer-polymer contact $V = V_{pp} + V_{ss} - 2\,V_{ps} = 2\,kT\chi la^2$ (χ is an interaction parameter of Huggins).

Due to assumptions made in deriving Eq. (62), the latter is correct only if the number of cross-linkages is not very small. Equation (62) can be solved graphically, and the dependence of the radius of gyration of the cross-linked chain on the average number of the cross-linkages can be found for every given value of the structure and thermodynamic parameters. The authors of [72] solved this equation for polystyrene which is cross-linked with diisocyanates (all parameters for this system are known and can be taken from the literature).

In deriving Eq. (62) it was supposed that a considerable gain in energy and a considerable decrease of dimensions take place during cross-linking. But the calcula-tion performed for polystyrene showed that the decrease of dimensions is not very great. This fact necessitated the introduction of the correction parameter in to Eq. (62), although it is difficult to evaluate the degree of this correction.

Simplifying Eq. (62) for a small number of cross-linkages, one obtains

$$\overline{R}_m^2 = \overline{R}_0^2/(m+1) \qquad (63)$$

But this simplification is not completely correct because of the assumption of a large number of cross-linkages in the derivation of Eq. (62).

Gordon et al.[73] proposed another empirical equation also based on thermo-dynamic considerations:

$$\overline{R}_m^2 = \overline{R}_0^2/(m + 1)^q \tag{64}$$

where $q \sim 0.2$.

Evidently this relation is more probable for the chains with fewer cross-linkages. Later on Ross-Murphy confirmed the validity of this relation using Monte Carlo calculations[74].

Allen et al.[72] compared the results of their calculations with experimental data measuring the intrinsic viscosity of intramolecularly cross-linked polystyrene. But their conclusion that these matched well is not completely convincing because of the inaccurate graphical presentation of their results. (The recalculation of $\overline{R}_m^2$ from the values of τ from Fig. 5 of the second paper[72] does not give the values presented in the same figure). So the accuracy of this theory and the possibility of its application to the interpretation of experimental data require further study.

The approach considered above is the only attempt at an analytical calculation of dimensions of the chains with intramolecular cross-linkages. Another approach is a simulation of this process using the Monte Carlo method[74-78].

Bonezkaya, Elyashevitch et al.[75] simulated so-called momentary cross-linking when the conformation of the chain is not changed until the end of the reaction. Macromolecules were simulated by a random walk procedure on the tetrahedral lattice assuming that the chain can be cross-linked only at sites of self-intersection. The authors[75] assumed that the probability of the reaction between two definite units is proportional to the number of conformations in which these units are close one to another. Then the population of the chains with a given number of cross-linkages can be obtained from the population of non-cross-linked conformations. The mean-square radius of gyration for the chain with m cross-linkages can be calculated as follows:

$$\overline{R}_m^2 = \left(\sum_{n_i \geqslant m} C_{n_i}^m q^{n_i - m} R_i^2 \right) \bigg/ \left(\sum_{n_i \geqslant m} C_{n_i}^m q^{n_i - m} \right) \tag{65}$$

where n_i is the number of units drawn together in a given conformation (with summaration over all conformations in which the number of pairs drawn together is not less than the number of cross-linkages); $C_{n_i}^m$ is the number of ways of cross-linking m pairs out of n_i possible pairs; q is the statistical weight of pairs which have been brought close to each other but not cross-linked.

During the computation, a number of non-cross-linked conformations were chosen at random, and the dimensions were evaluated approximately, using (65).

The results showed the monotonic decrease of chain dimensions during cross-linking (Fig. 7). The momentary cross-linking apparently corresponds to "fast" reactions when the reaction time is of the same order as the time of the change of the conformation. One can assume that usual "slow" chemical reactions for which

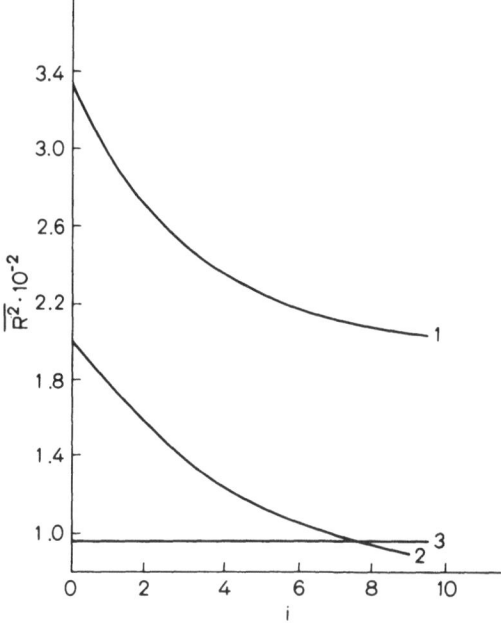

Fig. 7. Mean-square radius of gyration versus the number of cross-
linkages for the model of instanta-
neous cross-linking[75]

the time of reaction is much longer than the time in which the conformation changes
are accompanied by a more significant decrease of dimensions during cross-linking.
Such a case was considered by Platé, Noah et al.[76-78] who used Monte Carlo simulation.

In these papers, the chains were simulated on various lattices allowing self-
intersection. The consequent cross-linking of each particular chain was allowed, and
then the parameters obtained were averaged over the population of chains with a
given number of cross-linkages. The procedure of simulation included the following
stages: the random conformation was built; the number of reactive contacts was calculated (all non-cross-linked self-intersecting contacts were considered reactive); and
then one of the contacts was cross-linked with a probability:

$$W_j = \beta Z_{j-1} \tag{66}$$

where W_j is the probability of jth cross-linkage formation; Z_{j-1} is the number of
reactive contacts in the conformation with $j-1$ cross-linkages; and β is a normaliza-
tion coefficient.

If the chain was not cross-linked, one built the new conformation with the same
number of cross-linkages $(j-1)$ an so on up to the cross-linkage formation. The procedure was repeated until a given number of cross-linkages was formed. In Fig. 8,
the dependences of the relative average dimensions of polymer coils on the degree
of cross-linking are presented. It can be seen that chain dimensions essentially
decrease with cross-linking, and that this effect is greater with an increase of chain
length.

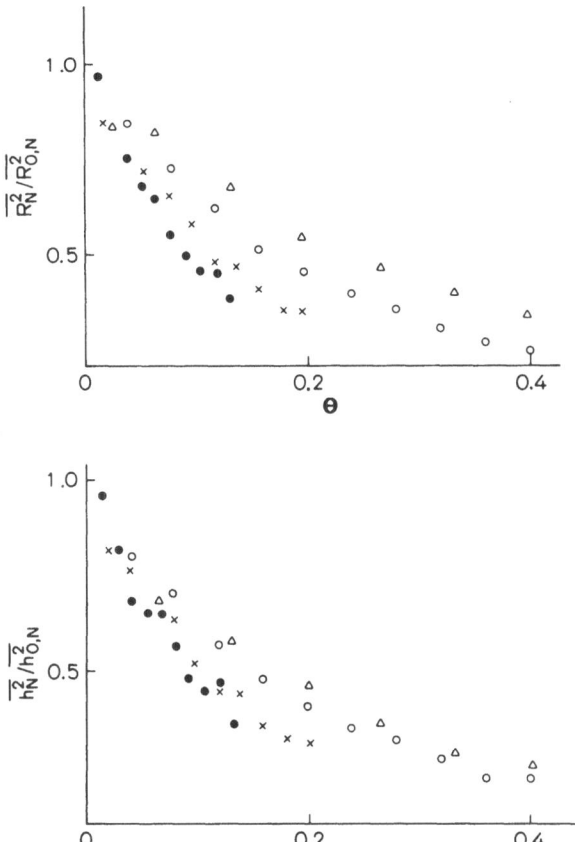

Fig. 8. Mean-square dimensions of cross-linked macromolecules versus the degree of cross-linking θ for model chains of 30 ($\triangle$), 50 ($\circ$), 100 ($\times$), 150 ($\bullet$) units [77]

2 Kinetics of Intramolecular Cross-Linking

The reaction rate corresponding to the model described above (the relatively slow chemical reaction between any two units of the macromolecule which come close to each other because of the flexibility of the chain) must be proportional to the number of reactive contacts in a partially cross-linked coil Z_j (j is the number of cross-linkages). Assuming that an average number of contacts is independent of the cross-linkage configuration [77], one can write the following system of kinetic equations [76–77]:

$$dC_0/dt = -k_0 \bar{Z}_0 C_0$$
$$dC_j/dt = k_0 (\bar{Z}_{j-1} C_{j-1} - \bar{Z}_j C_j) \qquad (67)$$
$$dC_M/dt = k_0 \bar{Z}_{M-1} C_{M-1}$$

where C_j is the number of chains with j cross-linkages; M is the maximum number of cross-linkages, and k_0 is the rate constant of elementary cross-linking.

Then the average number of cross-linkages in the chain

$$\bar{n}(t) = C^{-1} \sum_{j=1}^{M} j C_j(t)$$

(where $C = \sum_{j=0}^{M} C_j$ is the total number of chains) is determined as follows:

$$d\bar{n}/dt = k_0 C^{-1} \sum_{j=0}^{M-1} \bar{Z}_j C_j \tag{68}$$

So the problem of calculating the kinetics of intramolecular cross-linking is to find equilibrium average values of Z_j. The analytical approach to the estimation of Z_j is impossible for the reasons mentioned above in connection with the calculation of chain dimensions.

Platé, Noah et al.[76-77] studied the kinetics of intramolecular cross-linking using the Monte Carlo method. The procedure of the simulation was analogous to that described above. The time between two consequent cross-linking was:

$$t = \beta(m + \xi)/k_0 \tag{69}$$

where m is a number of conformations built before the formation of the cross-linkage[78]; ξ is a random number evenly distributed between 0 and 1; the procedure was repeated up to given time, T_{max}.

In Fig. 9 the change of the number of contacts with an increase in the number of cross-linkages is shown. The increase of the number of contacts is due to the decrease of the effective volume of the polymer coil during cross-linking. The decrease of the number of contacts for large degrees of cross-linking can be explained by exhaustion of free reactive groups. The numerical solution of Eq. (67) with $\bar{Z}_j$ obtained from the computer experiment gives the kinetic curve of the reaction. In Fig. 10 the results of such calculations are compared with data obtained directly by simulation. The good agreement of the two approaches with the calculations confirms the validity of the assumption made when the deriving Eq. (67) that the average number of contacts is independent on the configuration of cross-linkages.

It can be seen from Fig. 9 that the initial part of the curve can be represented in linear form:

$$\bar{Z}_j = A_j + B \tag{70}$$

The substitution of Eq. (70) into Eq. (68) results in

$$d\bar{n}/dt = k_0(A\bar{n} + B) \tag{71}$$

The solution of Eq. (71) is

$$\bar{n}(t) = (B/A)[\exp(k_0 At) - 1] \tag{72}$$

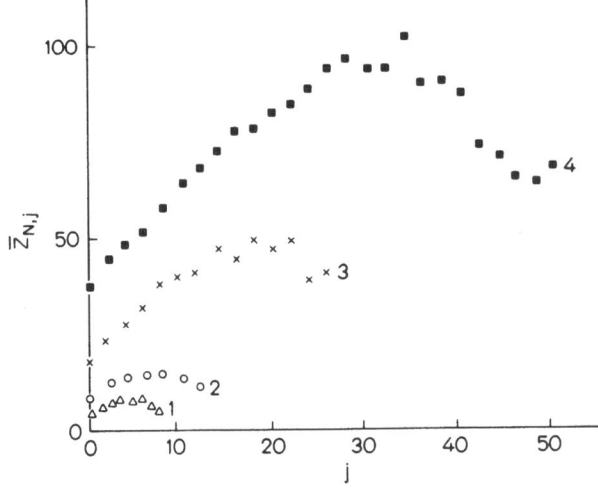

Fig. 9. Average number of reactive contacts $\bar{Z}_{N,j}$ versus the number of cross-linkages j; N = 30 (△), 50 (○), 100 (×), 150 (■)[77]

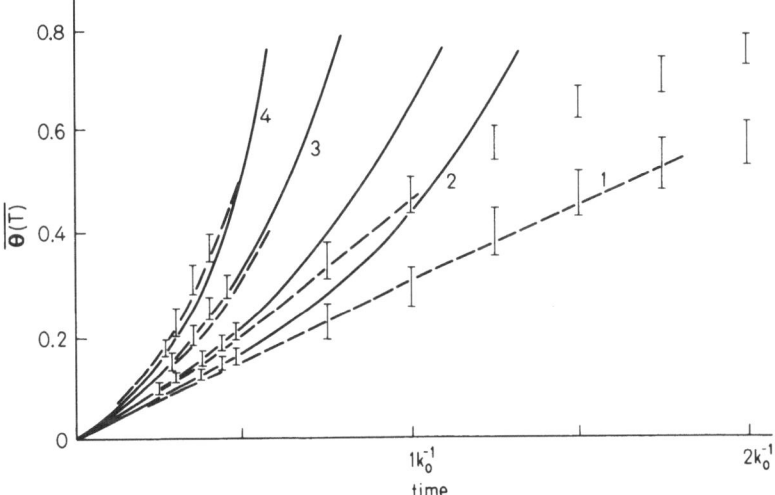

Fig. 10. Average degree of cross-linking $\bar{\theta}$ versus time; broken curves, the solution of Eqs. (67); solid curves, the solution of Eq. (72); points, Monte Carlo calculation for N = 30 (1), 50 (2), 100 (3), 150 (4)[77]

and with A and B found from Fig. 9, also agrees very well with the results of computer experiment. That is, the kinetics of intramolecular cross-linking can be initially described by linear approximation.

It should be pointed out in conclusion that intramolecular cross-linking is an autoaccelerated reaction, and that the initial rate and the degree of autoacceleration increase with an increase in chain length. Another important conclusion[76,77] is the existence of a uniform relationship between the kinetics of the reaction and the equilibrium properties of partially cross-linked chains.

3. Composition Heterogeneity of Cross-Linked Macromolecules

The composition distribution of the products for various degrees of cross-linking was also calculated in[77] using Monte Carlo simulation.

The exact distribution of the number of cross-linkages at any moment in time is determined by the solution of Eq. (67). The dispersion of this distribution is

$$D = \overline{n^2(t)} - [\overline{n}(t)]^2 \tag{73}$$

where $\overline{n^2(t)} = C^{-1} \sum\limits_{j=1}^{M} j^2 C_j(t)$ is the mean square number of cross-links in the chain.

If the average number of contacts in the chain were constant during the reaction, the process would be a random one. The number of cross-linkages would have a Poisson distribution with dispersion

$$D_P = \overline{Z}_0 k_0 t \tag{74}$$

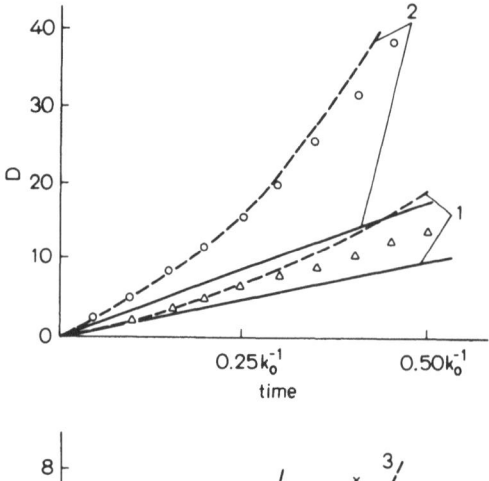

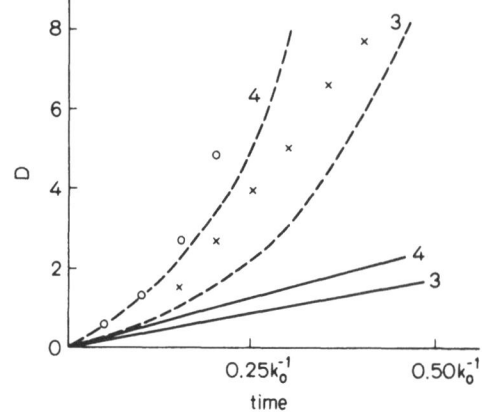

Fig. 11. Dispersion of cross-linkages number distribution versus time; solid curves, Poisson distribution; broken curves, linear approximation; points, Monte Carlo calculation for N = 30 (1), 50 (2), 100 (3), 150 (4)[77]

The calculation of the dispersion in the linear approximation gives the equation:

$$D_L = \exp(k_0At)(B/A)[\exp(k_0At) - 1] = \exp(k_0At)\,\bar{n}(t) \tag{75}$$

In Fig. 11 the values of the dispersion obtained from a computer experiment are compared with 'the results of calculations using Eqs. (74, 75). It can be seen from the figure that the true distribution is much wider than the Poisson distribution and that the width increases with an increase in chain length. In the initial stage, the dispersion is described well by the linear approximation. For short chains which are subjected to a reaction lasting some time the distribution becomes narrower due to the accumulation of chains with many cross-linkages (close to the maximum value) and the dispersion tends to that of a Poisson.

It should be pointed out in conclusion that the field dealing with intramolecular cross-linking reactions has not been well explored either experimentally or theoretically. The results of the few experimental studies[72, 79-82] unfortunately cannot yet be compared with the results of calculations. We hope that in the near future, intramolecular cross-linking reactions will attract the attention of chemists and that the study of this interesting and important domain will be intensified.

All the types of reactions of functional groups of macromolecules considered above were *kinetically controlled* reactions. We have not described *diffusionally controlled reactions* here. The theory of these reactions has developed rapidly in the last few years[83-89] and should be reviewed in a special paper.

C. References

1. Fettes, E. N. (ed.): Chemical reactions of polymers. New York: Interscience 1964
2. Morawetz, H.: Macromolecules in solutions. New York: Interscience 1976
3. Platé, N. A., Litmanovich, A. D., Noah, O. V.: Macromoleculyarnye reakzii. Moscow: Chimiya 1977
4. Platé, N. A. in: Kinetika i mekhanizm obrazovaniya i prevratscheniya macromolecul. Moscow: Nauka 1968
5. Platé, N. A., in: Uspekhi khimii i tekhnologii polimerov. Moscow: Chimiya 1971
6. Kemeny, J. G., Snell, J. L.: Finite Marcov chains. 1968
7. Fuoss, R. M., Watanabe, M., Coleman, B. D.: J. Polymer Sci *48*, 5 (1960)
8. Keller, J. B.: J. Chem. Phys. *37*, 2584 (1962)
9. Alfrey, T. Jr., Lloyd, W. G.: J. Chem. Phys. *38*, 318 (1963)
10. Arends, C. B.: J. Chem. Phys. *38*, 322 (1963)
11. Keller, J. B.: J. Chem. Phys. *38*, 325 (1963)
12. Lazare, L.: J. Chem. Phys. *38*, 727 (1963)
13. McQuarrie, D. A., McTague, J. P., Reiss, H.: Biopolymers *3*, 657 (1965); McQuarrie, D. A.: J. Appl. Prob. *4*, 413 (1967)
14. Mityushin, L. G.: Problemy peredatchi informatzii *9*, 81 (1973)
15. Stauff, J.: Z. Elektrochem. *48*, 550 (1942).
16. Noah, O. V., Litmanovich, A. D.: Vysokomolek. Soed. *A 19*, 1211, 1977
17. Silberberg, A., Simha, R.: Biopolymers *6*, 479 (1968)
18. Simha, R., Lacombe, R. H.: J. Chem. Phys. *55*, 2936 (1971)

19. Rabinovitz, P., Silberberg, A., Simha, R., Loftus, E., in: Stochastic prosses in chemical physics. Shuler, K. E. (ed.). New York: Interscience 1969
20. Silberberg, A., Simha, R.: Macromolecules 5, 332 (1972)
21. Lacombe, R. H., Simha, R.: J. Chem. Phys. 58, 1043 (1973)
22. Lacombe, R. H., Simha, R.: J. Chem. Phys. 61, 1899 (1974)
23. Vainstein, E. F., Berlin, Al. Al., Entelis, S. G.: Vysokomol. Soed. B17, 835 (1975)
24. Krishnaswami, P., Vadav, D. P.: J. Appl. Polymer Sci. 20, 1175 (1976)
25. Dobrushin, R. L.: Problemy peredatchi informatzii 7, 57 (1971)
26. Platé, N. A., Litmanovich, A. D.: 23rd International Congress of Pure and Applied Chemistry, Boston 1971; Preprints 8, 123 (1971)
27. Noah, O. V., Toom, A. L., Vasilyev, N. B., Litmanovich, A. D., Platé, N. A.: Vysokomol. Soed. A15, 877 (1973)
28. Platé, N. A., Litmanovich, A. D., Noah, O. V., Toom, A. L., Vasilyev, N. B.: J. Polymer Sci., Polymer Chem. Ed. 12, 2165 (1974)
29. Klesper, E., Gronski, W., Barth, V.: Makromol. Chem. 150, 223 (1971)
30. Klesper, E., Johnsen, A., Gronski, W.: Makromol. Chem. 160, 167 (1972)
31. Platé, N. A., Seifert, T., Stroganov, L. B., Noah, O. V.: Dokl. Akad. Nauk SSSR 223, 396 (1975)
32. Litmanovich, A. D. et al.: Vysokomol. Soed. A17, 1112 (1975)
33. Klesper, E., Barth, V., Johnsen, A.: 23rd International Congress of Pure and Applied Chemistry, Boston, 1971; Preprints 8, 151 (1971)
34. Berlin, Al. Al., Vainstein, E. F., Entelis, S. G.: Vysokomol. Soed. B20, 275 (1978)
35. Platé, N. A., Litmanovich, A. D., Noah, O. V., Golyakov, V. I.: Vysokomol. Soed. A11, 2204 (1969)
36. Litmanovich, A. D., Platé, N. A., Noah, O. V., Golyakov, V. I.: Europ. Polymer J.-Suppl. 1969, 517
37. Noah, O. V., Toom, A. L., Vasilyev, N. B., Litmanovich, A. D., Platé, N. A.: Vysokomol. Soed. A16, 412 (1974)
38. Noah, O. V., Litmanovich, A. D., Platé, N. A.: J. Polymer Sci., Polymer Phys. Ed. 12, 1711 (1974)
39. Frensdorf, H. K., Ekiner, O.: J. Polymer Sci., A2, 1157 (1967)
40. Platé, N. A., Noah, O. V.: Vysokomol. Soed. B19, 483 (1977)
41. Kuchanov, S. I., Brun, E. B.: Dokl. Akad. Nauk SSSR 227, 662 (1976)
42. Noah, O. V., Torchilin, V. P., Litmanovich, A. D., Platé, N. A.: Vysokomol. Soed. A16, 668 (1974)
43. Krenzel, L. B., Litmanovich, A. D.: Vysokomol. Soed. B16, 372 (1974)
44. Pismen, L. M.: Vysokomol. Soed. A14, 1861 (1972)
45. Flory, P. J.: J. Am. Chem. Soc. 61, 1518 (1939)
46. Cohen, E. R., Reiss, H.: J. Chem. Phys. 38, 680 (1963)
47. Gordon, M., Hillier, J. H.: J. Chem. Phys. 38, 1376 (1963)
48. Lee, D. F., Scanlan, J., Watson, W. F.: Proc. Roy. Soc. A-273, 345 (1963)
49. McQuiston, R. B., Lichtman, D.: J. Math. Phys. 9, 1680 (1968)
50. Lewis, C. W.: J. Polymer. Sci. PA-2, 10, 377 (1972)
51. Barron, T. H. K., Boucher, E. A.: Trans. Faraday Soc. 65, 3301 (1969); Trans. Faraday Soc. 66, 2320 (1970)
52. Boucher, E. A.: J. Chem. Soc., Faraday Trans., I, 68, 2295 (1972)
53. Boucher, E. A.: J. Chem. Phys. 59, 3848 (1973)
54. Boucher, E. A.: Chem. Phys. Letters 17, 221 (1972); J. Chem. Soc., Faraday Trans., 2, 69, 1839 (1973)
55. Boucher, E. A.: Makromol. Chem. 173, 253 (1973)
56. Barron, T. H. K., Bawden, R. J., Boucher, E. A.: J. Chem. Soc. Faraday Trans. 2, 70, 651 (1974)
57. Boucher, E. A.: J. Chem. Soc., Faraday Trans. 2, 72, 1697 (1976)
58. Gonzalez, J. J., Hemmer, P. C.: J. Polymer Sci., Polymer Letters Ed. 14, 645 (1976); J. Polymer Sci., Polymer Phys. Ed. 15, 321 (1977)

59. Morawetz, H., Goodman, N., Macromolecules *3*, 699 (1970)
60. Goodman, N., Morawetz, H.: J. Polymer Sci. *C-31*, 177 (1970)
61. Goodman, N., Morawetz, H.: J. Polymer Sci. PA-2, *9*, 1657 (1971)
62. Morawetz, H., Cho, J.-R., Gans, P. J., Macromolecules *6*, 624 (1973)
63. Morawetz, H.: Pure Appl. Chem. *38*, 267 (1974)
64. Sisido, M.: Macromolecules *4*, 737 (1971)
65. Sisido, M.: Polymer J. *3*, 84 (1972)
66. Sisido, M.: Seibutsu Butsuri *14*, 135 (1974)
67. Sisido, M., Mitamura, T., Imanishi, Y., Higashimura, T.: Macromolecules *9*, 316, 320 (1976); *10*, 125 (1977)
68. Sisido, M., Tamura, F., Imanishi, Y., Higashimura, T.: Biopolymers *16*, 2723 (1977)
69. Zimm, B. H., Stockmayer, W. H.: J. Chem. Phys. *17*, 1301 (1949)
70. Romantzova, I. I.: Dissertation, Moscow 1978
71. Edwards, S. F., Freed, K. F.: J. Phys. *C3*, 739, 750, 760 (1970)
72. Allen, G., Burgess, J., Edwards, S. F., Walsh, D. J.: Proc. Roy. Soc., London *A-334*, 453, 465, 477 (1973)
73. Gordon, M., Torkington, J. A., Ross-Murphy, S. B.: Macromolecules *10*, 1090 (1977)
74. Ross-Murphy, S. B.: Polymer *19*, 497 (1978)
75. Bonezkaya, N. K., Irzhak, V. I., Elyashevitch, A. M., Enikilopyan, N. S.: Dokl. Akad. Nauk SSSR *222*, 140 (1975)
76. Romantzova, I. I., Taran, Yu. A., Noah, O. V., Platé, N. A.: Dokl. Akad. Nauk SSSR *234*, 109 (1977)
77. Romantzova, I. I., Taran, Yu. A., Noah, O. V., Elyashevitch, A. M., Gotlib, Yu. Ya., Platé, N. A.: Vysokomol. Soed. *A 19*, 2800 (1977)
78. Elyashevitch, A. M.: Vysokomol. Soed. *A 20*, 951 (1978)
79. Irzhak, V. I., Kuzub, L. I., Enikolopyan, N. S., in: Sintez i fiziko-khimiya polimerov. 1973, is. 12
80. Irzhak, V. I., Kuzub, L. I., Enikolopyan, N. S.: Dokl. Akad. Nauk SSSR *214*, 1340 (1974)
81. Kuzub, L. I., Irzhak, V. I., Bogdanova, L. M., Enikolopyan, N. S.: Vysokomol. Soed. *B16*, 431 (1974)
82. Raspopova, E. N., Bogdanova, L. M., Irzhak, V. I., Enikolopyan, N. S.: Vysokomol. Soed. *B16*, 434 (1974)
83. Wilemski, G., Fixman, M.: J. Chem. Phys. *58*, 4009 (1973)
84. Doi, M.: Chem. Phys. *9*, 455 (1975)
85. Doi, M.: Chem. Phys. *11*, 107, 115 (1975)
86. Sunagawa, S., Doi, M.: Polymer J. *7*, 604 (1975)
87. Sakata, M., Doi, M.: Polymer J. *8*, 409 (1976)
88. Sunagawa, S., Doi, M.: Polymer J. *8*, 239 (1976)
89. Kozlov, S. V.: Vysokomol. Soed. *B18*, 642 (1976)

Received September 25, 1978
H.-J. Cantow (editor)

Author Index Volumes 1–31

Polymers

Properties and Applications

Editorial Board:
H.-J. Cantow, H. J. Harwood,
J. P. Kennedey, A. Ledwith,
J. Meißner, S. Okamura,
G. Olivé, S. Olivé

Springer-Verlag
Berlin
Heidelberg
New York

Volume 1
B. Rånby, J. F. Rabek

ESR Spectroscopy in Polymer Research

1977. 356 figures, 29 tables. XIV, 410 pages
ISBN 3-540-08151-8

Contents:
Generation of Free Radicals. – Principles of ESR Spectroscopy. – Experimental Instrumentation of Electron Spin Resonance. – ESR Study of Polymerization Processes. – ESR Study of Degradation Processes in Polymers. – ESR Study of Polymers in Reactive Gases. – ESR Studies of the Oxidation of Polymers. – ESR Studies of Molecular Fracture in Polymers. – ESR Studies of Graft Polymerization. – ESR Studies of Crosslinking. – Application of Stable Free Radicals in Polymer Research. – ESR Spectroscopy of Stable Polymer Radicals, Polyradical Anions and their Low Molecular Analogues. – ESR Study of Ion-Exchange Resins. – References. – Index.

Volume 2
H.-H. Kausch

Polymer Fracture

1978. 180 figures. X, 332 pages
ISBN 3-540-08786-9

Contents:
Deformation and Fracture of High Polymers, Definition and Scope of Treatment. – Structure and Deformation. – Statistical, Continuum Mechanical, and Rate Process Theories of Fracture. – Strength of Primary Bonds. – Mechanical Excitation and Scission of a Chain. – Identification of ESR Spectra of Mechanically Formed Free Radicals. – Phenomenology of Free Radical Formation and of Relevant Radical Reactions (Dependence on Strain, Time, and Sample Treatment). – The Role of Chain Scission in Homogeneous Deformation and Fracture. – Molecular Chains in Heterogeneous Fracture.

Volume 3
A. Knop, W. Scheib

Chemistry and Application of Phenolic Resins

1978. 111 figures, 87 tables. Approx. 280 pages
ISBN 3-540-09051-7

Contents:
Historical and Economical Development of Phenolic Resins. – Raw Materials. – Reaction Mechanisms. – Resin Production. – Physiology and Environmental Protection. – Analytic Methods. – Degradation of Phenolic Resins by Heat, Oxygen and High Energy Radiation. – Modified and Thermal-Resistant Resins. – Composite Wood Materials. – Molding Compounds. – Heat and Sound Insulation Materials. – Industrial Laminates and Paper Impregnation. – Coatings. – Foundry Resins. – Abrasive Materials. – Friction Materials. – Phenolic Resins in Rubbers and Adhesives. – Phenolic Antioxidants. – Other Applications. – Index.

Polymer Bulletin

Editors:

Prof. H.-J. Cantow
Institute of Macromolecular
Chemistry
University of Freiburg
Stefan-Meier-Strasse 31
D-78 Freiburg/Germany

Prof. J.P. Kennedy
Dept. of Polymer Science
The University of Akron
Akron, OH 44325/USA

Prof. T. Saegusa
Dept. of Synthetic Chemistry
Kyoto University
Kyoto, 606 Japan

The articles are to be sent to one of the editors or to
Springer-Verlag Berlin Heidelberg New York

Polymer Bulletin

Preface

To cope with the rapid progress of polymer science, a new
journal is now published characterized by emphasis on rapid
publication of papers containing a most concise description of
results.

The character of the new journal is between the purely archival
journal of full papers and the so-called "letter journals" consisting exclusively of short communications.

The journal consists of one volume a year, published in 12
issues.

Subscription information upon request.

Springer
International